Manual of medical entomology

Fourth edition

Manual of
medical entomology
Fourth edition

DEANE P. FURMAN
University of California, Berkeley
and
E. PAUL CATTS
Washington State University

Cambridge University Press

Cambridge
London   New York   New Rochelle
Melbourne   Sydney

Published by the Press Syndicate of the University of Cambridge
The Pitt Building, Trumpington Street, Cambridge CB2 1RP
32 East 57th Street, New York, NY 10022, USA
296 Beaconsfield Parade, Middle Park, Melbourne 3206, Australia

Fourth edition © Cambridge University Press 1982
Third edition first published by Mayfield Publishing Company 1970
Third edition first published by Cambridge University Press 1980

Printed in the United States of America

Library of Congress Cataloging in Publication Data
Furman, Deane Philip, 1915–
Manual of medical entomology.

1. Arthropod vectors.   2. Arthropoda –
Identification.   I. Catts, E. Paul   II. Title.
III. Title: Medical entomology.
RA641.A7F87   1982   614.4′32   81–10105
ISBN 0 521 29920 9   AACR2

48,498

Contents

Preface

With the fourth edition of the *Manual of Medical Entomology,* we have expanded the taxonomic keys to encompass a broader geographic scope of arthropods of medical or veterinary importance. This manual had its beginning as an assemblage of mimeographed keys used by students in the medical entomology course at the University of California, Berkeley. For that reason the early editions emphasized the fauna of California. In addition to including more of the common medically important species of North America, with this edition we have expanded coverage to include genera of mosquitoes of the Western Hemisphere and genera of the kissing bugs of the world.

Users of earlier editions will recognize many of the illustrations used here, but nearly all have been redrawn, and more than 200 new ones have been added to augment new key sections or to illustrate poorly understood characters.

The improved quality of this revision is due in large measure to generous contributions and constructive suggestions made by colleagues and specialists of the various arthropod groups. We acknowledge this help with sincere gratitude. Our special thanks go to Roger D. Akre, Richard F. Darsie, K. C. Emerson, Stephanie Clark Gil, K. C. Kim, Robert W. Lake, John F. MacDonald, Thomas N. Mather, Bernard C. Nelson, Harold E. Stark, William J. Turner, George Uetz, and Robert K. Washino.

Finally our appreciation goes to those former students in our classes of medical entomology at the University of California, Berkeley, at the University of Delaware, and at Washington State University, who through their questions and suggestions have helped in a grand way to improve this, our latest effort. It is to them and to the students yet to come that we dedicate this manual.

DPF
EPC

Introduction

Medical entomology is a biological science dealing with arthropods that affect the health of humans or other vertebrate animals, through their ability to transmit causal agents of disease, through their direct attacks, or as a result of the harmful effects produced by contact. This defines medical entomology in the broad sense to include both public health and veterinary entomology.

This manual is primarily a teaching tool to familiarize students with the use of taxonomic keys for identifying medically important arthropods and with those characters of taxonomic value for each of the diverse groups of arthropods. Accurate identification is essential to any definitive diagnosis of an arthropod infestation and hence to valid recommendations for its control. The second purpose of this manual is to serve as a concise reference collection of keys for the medical entomologist as the first step in identification of medically important arthropods before more specific keys are employed. These exercises then are largely in the form of taxonomic keys which, with the help of illustrations, are designed as studies in diagnosis and not in systematic entomology. Additional techniques are described for the study of arthropods.

Initial study is made of a generalized arthropod, followed by brief, intensive consideration of several lines of development undergone by arthropod mouthparts, inasmuch as structural variations from the generalized types not only provide guideposts to identification, but also serve as important clues to function of the organ and habits of the organism. Techniques of rearing, collecting, preserving, and mounting arthropods for study are covered. Proper dissection of arthropods and methods of identifying a blood pathogen and the source of a blood meal are described.

In using the syllabus it is necessary that the student read the exercise before reporting to the laboratory, correlating with it information gained from the textbook and collateral reading in reference works. Very few drawings are required because most of the more difficult key characters have been illustrated. However, the conscientious student should add labeled sketches of his own whenever questionable characters appear with the specimens provided. In this way structures can be clarified that otherwise might be misinterpreted if based only upon a description.

In addition to the formal laboratory exercises each student should prepare one or more special reports detailing the life history of these subject arthropods. If possible such arthropods should be collected from their natural habitats and reared so that preserved, mounted, and labeled specimens will serve as part of the report. The report also may include literature research of the original taxonomic description of the arthropod to emphasize those specific characters that distinguish it from all other species. The student may need to prepare his own original descriptions for life stages that have no published descriptions.

References

Ebeling, W. 1975. *Urban Entomology*. University of California, Division of Agricultural Sciences, Berkeley.

Faust, E. C., P. C. Beaver, and R. C. Jung. 1975. *Animal Agents and Vectors of Human Disease*, 4th ed. Lea & Febiger, Philadelphia.

Faust, E. C., P. F. Russell, and R. C. Jung. 1970. *Craig and Faust's Clinical Parasitology*, 8th ed. Lea & Febiger, Philadelphia.

Gordon, R. M., and M. M. J. Lavoipierre. 1962. *Entomology for Students of Medicine*. Blackwell, Oxford.

Harwood, R. F., and M. T. James. 1979. *Entomology in Human and Animal Health*, 7th ed. Macmillan, New York. 548 pp.

Horsfall, W. R. 1962. *Medical Entomology*. Ronald Press, New York.

Leclercq, M. 1969. *Entomological Parasitology: The Relations Between Entomology and the Medical Sciences*. Pergamon Press, New York.

Roberts, F. H. S. 1952. *Insects Affecting Livestock*. Angus & Robertson, London.

Smith, K. G. V. (ed.). 1973. *Insects and Other Arthropods of Medical Significance*. Pub. 720. British Museum (Natural History), London. 561 pp.

Soulsby, E. J. L. 1968. *Helminths, Arthropods and Protozoa of Domesticated Animals*. Williams & Wilkins, Baltimore.

Southcott, R. V. 1976. Arachnidism and allied syndromes in the Australian region. *Rec. Adelaide Children's Hosp. 1*:97–186.

Wigglesworth, V. B. 1964. *The Life of Insects*. Weidenfeld & Nicolson, London.

1   Insect morphology

The vast majority of all known kinds of animals (72%) are insects. Each kind, or species, of insect has its own ecological niche requirements, and no two species have the same set either of requirements or of behavioral traits. These differences translate into an array of morphological modifications that better adapt a species to its niche. Such morphological differences can be used to distinguish one species from another. The correct identification of a medically important insect or other arthropod is the primary step in accomplishing its control or elimination.

Identification of the myriad insects and other arthropods that are of medical importance requires that students be able to compare and visually discriminate among a wide range of structural characteristics that differ among species. By the use of dichotomous keys, comparisons are made sequentially to lead a student or researcher through a series of correct descriptive choices that best matches the specimen at hand.

In order to use the keys, the student or researcher must be familiar with the structural terminology that describes the generalized insect. Many such terms are used throughout the keys in this manual regardless of the group being studied. In addition, there are a few terms used only for structures peculiar to certain groups of insects (e.g., ectospermalege in Cimicidae; trumpet and air tube in mosquito pupae and larvae, respectively). The student must learn these peculiarities in order to become familiar with the use of keys for various groups important to the medical entomologist.

As a starting point, this exercise will introduce terms for structures that make up any insect. This knowledge will equip you, the student, to practice using dichotomous keys as an aid to identifying unknown, medically important insects. This knowledge also will prepare you to make comparative judgments of insect morphology – that is, a functional interpretation of advantages or disadvantages of structural modifications or specializations in medically important insects.

A generalized insect: the cockroach

Study the cockroach – any house-frequenting species will do – as an example of a generalized insect of minor medical importance. Cockroaches are modestly specialized with respect to their running capabilities, or cursorial habits, and their structure has changed little in the last 350 million years. Using the following description and Figure 1.1 as a guide, locate those structures that are visible on your cockroach specimen.

The adult insect body is divided into three functional regions: **head** (the nerve and sensory control center and the food-ingesting unit), **thorax** (the locomotor unit), and **abdomen** (the food-digesting, food-assimilating, and reproductive unit).

Head
The head bears the mouthparts (with their associated appendages as palpi), one pair of antennae, the ocelli (if present), and the compound eyes. Chewing-type mouthparts, as in the cockroach, are considered to represent a relatively primitive or generalized type. Other insects demonstrate more specialized types: piercing-sucking and sponging. Mouthparts will be considered in more detail in Exercise 2.

The six segments that make up the head are largely fused with each other, so there are few external morphological landmarks that give clues to the limits of distinct segments. Thus most subdivisions of the head are designated as regions or areas. The clypeal region of the head is the area to which the labrum or front "lip" is appended. The occipital head region is the back of the head capsule, and the ocellar, or vertex, region is between the eyes. The genal regions are the sides of the head and the cervical region is the neck. The stomal, or oral, region includes the mouthparts and the supporting margins of the head capsule.

Thorax
The thorax is composed of three segments. Proceeding from front to rear (anterior to posterior, or cephalad to caudad), they are the **prothorax, mesothorax,** and **metathorax.** Each of these segments bears

one pair of legs. The legs are divided into distinct sections, as the name **arthropod** (jointed leg) implies. The small, proximal segment – that one attached to the body – is the **coxa.** Moving distally (toward the extremity of the leg), the segments in order are **trochanter** (also a small segment), **femur, tibia,** and **tarsus,** the larger leg segments. The tarsus is usually subdivided into a number of tarsal segments, characteristically five, and terminates in a **pretarsus.** The pretarsus (Figure 1.2) consists of a central **arolium,** typically padlike; a pair of **claws** (ungues); and an **unguitractor plate.** The arolium may be modified into a spinelike process, the **empodium,** and under each claw there may be a straplike or padlike **pulvillus.**

The meso- and metathorax each possesses a pair of wings except for very primitive insects or specialized insects that have secondarily lost their wings during the course of evolutionary changes. The **wings** are formed from membranous expansions of the thoracic body wall that are reinforced or strengthened by tubular thickenings called **veins.** The form and location of specific veins are important characters for identification of many flying insects.

Many groups of insects (orders) possess both pairs of wings, which function in the characteristic mode of lift surfaces moved in flight. In some groups, wings also may be modified for other functions such as shields to protect the fragile flight wings (e.g., cockroaches and beetles) or as gyroscopic balances used in flight (e.g., flies).

The meso- and metathorax also each typically bears a pair of lateral spiracles (respiratory apertures).

Thoracic segments are made up of a number of hardened plates, or **sclerites.** The dorsal sclerite is called the **notum,** the ventral sclerite is the **sternum,** and the lateral sclerites are **pleura.** A pleuron may be subdivided into smaller units with names such as **mesepimeron, metepisternum, hypopleuron,** and **meron.** These details will be treated more fully in those exercises where pleurites are used as key characters.

Abdomen

The abdomen of more primitive insects usually is composed of 11 segments, but in higher groups normally only 9 or 10 are visible. The terminal segments may coalesce or telescope so that only 3 or 4 can be seen without manipulating the specimen. The dorsal sclerite of each segment is the **tergum** (or tergite) and the ventral sclerite the **sternum** (sternite). If distinct lateral plates are present, as in the sucking lice, they are called **pleura,** or pleurites. Each pleurite, or the pleural region of the tergite, usually surrounds the external respiratory opening, the **spiracle.**

The caudal (posterior) region of the abdomen typically bears the external **genitalia** and associated structures and terminates with the **anus.** Male genitalia tend to be more elaborate and conspicuous, taking the form of claspers, rods, and an intromittent apparatus. Female genitalia are simpler, often valvelike or slitlike, and may be telescoped to be quite hidden. In a few groups the female ovipositor is elongated into a conspicuous structure for subsurface egg depositing or as a defensive weapon (e.g., bees, wasps).

Comparison of cockroach with a higher insect

After completing study of the cockroach, examine a higher insect (such as a blow fly or a house fly) and locate the structures homologous to those of the cockroach that occur on that insect. Use Figure 1.3 to aid in making this comparison. Note the exaggerated morphological dominance of the mesothorax, reflecting the more highly developed capabilities of flight in flies.

If time is available, further detailed comparisons of sclerite homologies can be made within the Diptera using the thorax of the mosquito, horse fly, or house fly.

References

See those listed for Exercise 2.

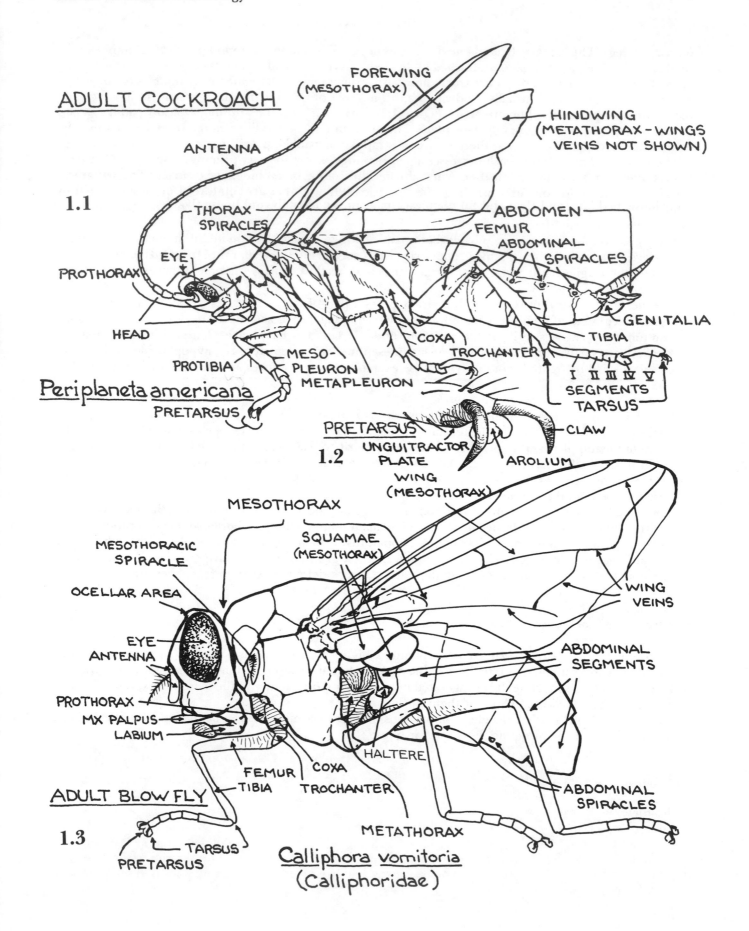

ADULT COCKROACH

FOREWING
(MESOTHORAX)

HINDWING
(METATHORAX - WINGS
VEINS NOT SHOWN)

ANTENNA

1.1

THORAX
SPIRACLES

ABDOMEN
FEMUR
ABDOMINAL
SPIRACLES

EYE

PROTHORAX

GENITALIA

HEAD

COXA

TIBIA

TROCHANTER

PROTIBIA

MESO-
PLEURON
METAPLEURON

I II III IV V
SEGMENTS
TARSUS

Periplaneta americana

PRETARSUS

PRETARSUS

CLAW

1.2

UNGUITRACTOR
PLATE

AROLIUM

WING
(MESOTHORAX)

MESOTHORAX

SQUAMAE
(MESOTHORAX)

MESOTHORACIC
SPIRACLE

WING
VEINS

OCELLAR AREA

EYE
ANTENNA

ABDOMINAL
SEGMENTS

PROTHORAX
MX PALPUS
LABIUM

HALTERE

FEMUR
TIBIA

COXA
TROCHANTER

ABDOMINAL
SPIRACLES

ADULT BLOW FLY

METATHORAX

1.3

TARSUS

PRETARSUS

Calliphora vomitoria
(Calliphoridae)

2   Mouthparts

Insect mouthparts may be divided into two general categories: (1) chewing or mandibulate, as found in the cockroach, beetle, and bird louse; and (2) sucking or haustellate, as found in the bed bug, mosquito, house fly, flea, and butterfly. Sucking mouthparts can be subdivided further into (a) piercing, as in the mosquito, body louse, bed bug, and flea; and (b) nonpiercing, as in the blow fly and house fly.

The basic preoral (before the true mouth) structures of all insects have evolved from the type used in a generalized chewing function, with the **mandibles** as the principal tearing and probing structures. Sucking mouthparts are modified greatly from the chewing type, often with the elongated **maxillae, mandibles,** and variously modified **labium** as the principal structures. Most medically important insects belong to the category with sucking mouthparts.

Chewing insects typically feed on solid substrates, as opposed to sucking insects, which feed on various fluids. As vectors of pathogens, chewing insects are of far less importance than sucking insects to the medical entomologist, but study of chewing insects' mouthparts is needed as reference for understanding the origin and functioning of those highly modified structures seen in sucking insects. Chewing insects can serve as intermediate hosts of helminth parasites of vertebrates and as mechanical vectors of disease agents, but they do not introduce pathogenic agents into deeper tissues while feeding. Sucking insects, however, are able to introduce pathogens into subsurface tissues, including the bloodstream, and thus are of major importance as vectors of disease-causing agents.

Basically, insect mouthparts consist of upper (**labrum**) and lower (**labium**) lips separated by two pairs of lateral jaws (**mandibles** and **maxillae**), which embrace a central tonguelike structure (**hypopharynx**). The proximal anterior surface of the hypopharynx closes the back of the **true mouth,** and its posterior surface marks the location of the salivary pore. The front of the true mouth is closed by the posterior wall of the labrum, the **epipharynx.** Among different groups of insects, different oral structures are modified or exaggerated to be used as principal food-getting mouthparts. The labrum (or labroepipharynx) and hypopharynx are integral outgrowths of the cranial wall, whereas the mandibles, maxillae, and labium evolved from segmented appendages of the fused cephalic segments and are homologous to the thoracic legs.

Basic chewing-type mouthparts (Figure 2.1A–F)

Chewing mouthparts occur in cockroaches, locusts, crickets, beetles, and immatures of most holometabolous insects. Examine the head of a cockroach, and locate and manipulate the oral structures identified in Figure 2.1A–F. Dissect the mouthparts free from the head beginning with the labrum and proceeding to the labium while referring to the following description of preoral structures.

Labrum

This is a flaplike, sensory dorsal lip; there are gustatory areas on the ventral, or epipharyngeal, surface. The labrum is bounded above by the **clypeus,** but hangs free below. The labrum functions to keep food in the preoral cavity and to protect the basal articulations of the mandibles.

Mandibles

These are paired, darkly pigmented jawlike structures exposed upon removal of the labrum. They are toothed on the inner, or mesal, aspect and serve as the main scratching, grasping, cutting, and masticating structures. They articulate in a transverse plane, pivoting at two points located on the rim of the cranium. The cutting areas are distal, whereas the molar areas are proximal; these areas may be used separately or simultaneously. The basal mandibular brushes help align food.

Maxillae

Removal of the mandibles reveals the paired maxillae. Each is a complex of three distal units (**galea, lacinia,** and a segmented **palpus**) supported basally by the **cardo** and **stipes.** The lateral flaplike galea partly covers the outer face of the mandible and functions as a lateral lip closing the sides of the preoral cavity. The lacinia is distally toothed (maxadentes) and manipulates all food pieces cut by the mandibles. The maxillae open and close and rock back and forth simultaneously (on a single point of articulation on the cardo) in opposite phase to the mandibles. They push cut pieces of food into the preoral cavity for mastication in mandibular molar areas; they are used for secondary mastication of soft foods and serve to hold the food bolus in the preoral cavity from below. The lacinarasta (brushes) are probably tactile. The paired maxillary palpi function as sensory structures to locate food near the head and drum on food and feeding surface. The palpi determine food alignment (but do not manipulate food) in the preoral cavity and possess gustatory areas.

Hypopharynx

This is a fleshy lobe or tonguelike structure protruding into the preoral cavity between the bases of the maxillae. The salivary duct opens at its ventral base. The hypopharynx helps mix salivary secretions with the food bolus in the preoral cavity. It is mainly gustatory in function. Retraction of the hypopharynx helps draw the bolus from the preoral cavity through the **true mouth** and into the pharynx.

Labium

This is a complex, hinged flap that acts as the lower lip and resembles a pair of medially fused maxillae. The labium forms the floor, or back, of the preoral cavity and consists of a **mentum** base subdivided into smaller units, paired segmented **palpi,** and distal lobes, the **glossae** and **paraglossae.** The labium is a sensory structure, seldom functioning to align food. The paraglossae are displaced laterally during feeding as the labium pushes its salivary wetted anterior surface against the food bolus.

Observation of chewing operation

Operation of chewing mouthparts can be observed in the following way.

Starve and withhold water from a cockroach for 24 hours, and fasten the live specimen, belly up, in a dish of warm paraffin. Hold the insect in place until the paraffin cools and hardens. Under a stereoscopic microscope offer the immobilized insect small amounts of banana, honey, or a sweet fluid (e.g., soft drink) on the tip of an applicator stick. Watch the movement and sequence of feeding by the battery of preoral structures. The feeding action can be summarized as follows: cutting and tearing pieces of food by mandibles; manipulation, and some mastication, by the maxilla; transport into the pharynx from the preoral cavity by the hypopharynx; the labrum, mandibles, galea, and paraglossae all functioning to confine the food to the preoral cavity. Even though chewing and sucking appear to describe distinct modes of ingestion, preoral structures of both types must also function in the common role of drinking or imbibing fluids.

Understanding the progressive transition and modification of piercing-sucking mouthparts from the basic chewing function is not as illogical or difficult as one might assume. The following types and subtypes are described to represent this transition, using arthropods of some medical importance.

Modified chewing-type mouthparts, mallophagan type (Figure 2.2)

The mouthparts of the more primitive bird lice (Mallophaga, Amblycera) show essentially the same arrangement of preoral structures as seen in the cockroach. The small size and depressed head form of the chewing louse require that the mouthparts be examined on a cleared, slide-mounted specimen. Study the mouthparts of an amblyceran under a compound microscope and locate those structures illustrated in the accompanying figures. The opposable **mandibles** are dark and heavily sclerotized. Exceptional differences from the basic chewing-type mouthparts are the picklike development of the **lacinial lobe** of the maxilla and the coordinated development and working of epipharynx and hypopharynx into a "crest" and "sitophore" to help shuttle feather barbules, or skin scrapings, into the pharynx (Buckup, 1959). The **galeal lobe** of the maxilla operates in a jawlike movement, but the lacinial lobe acts independently. The lacinia has a set of simple protractor retractor muscles to allow an in-and-out probing action by the thornlike lacinial tip.

Piercing mouthparts, hemipteran type (Figure 2.3)

The piercing-sucking apparatus of the hemipteran, or true bug, type is characterized by a fascicle of four, much elongated, thin, piercing stylets homologous with the mandibles and maxillae of the basic chewing mouthparts. These stylets are ensheathed in the large protective labium when not in use. Both labrum and hypopharynx are small and inconspicuous; they appear to have little direct function in feeding. Examine the proboscis or rostrum of an assassin bug (Reduviidae), and locate the structures shown in Figure 2.3.

In feeding by the triatomine, *Rhodnius prolixus,* the protracted labium makes contact with the feeding surface and usually ejects fluid from its tip, which quickly hardens to form a supporting feeding collar, or cone. The **labium** is shortened by telescoping and bending to give greater exposure to the stylets. The barbed mandibles probe with rapid alternating movements to draw and anchor the fascicle in the feeding surface. The **maxillae** are interlocked to form the dorsal food canal and salivary duct; they slide back and forth on each other to make deeper sampling thrusts and withdrawals (as deep as 1.5 mm) until a suitable blood vessel is penetrated near the entry site. In bed bugs the proboscis (labium and the contained stylets) is swung forward to about a 90° angle, and the body is flexed to produce a hump at the thorax; the bed bug rocks back and forth as the fascicle is thrust in and out of the feeding surface (Usinger, 1966).

Piercing mouthparts, anopluran type (Figure 2.4)

Anopluran mouthparts represent one extreme modification of the basic structures for a piercing function. In the Anoplura, the styletlike mouthparts are largely retracted internally into a ventrally located cephalic sac when not in use. Usually the labrum is the only visible mouthpart externally; it forms part of a short, eversible, snoutlike tube known as the haustellum, or mouth cone. Partial eversion of the haustellum exposes the buccal teeth, which are used to penetrate the outer, or horny, layer of host skin with a rapid action described as like a "rotary saw" (Lavoipierre, 1967). The completely embedded haustellum is then fully everted so that the buccal teeth anchor the head to the substrate.

The three piercing stylets, formed as a single bundle, are forced by muscle contraction from the cephalic sac out through the haustellum to probe for a capillary. Stylets usually probe 50–100 μm into tissues beyond the haustellum, but they are capable of extension triple that distance. The probing stylet bundle can be flexed from one side to the other as much as 45° and can be withdrawn very rapidly. The haustellum remains embedded for some time following withdrawal of the stylets.

The dorsal stylet is believed to be maxillary in origin, the median stylet is formed from the salivary duct, and the ventral stylet with serrated tips is formed from fused apical elements of the labium. There is no vestige of the mandibles remaining (Stojanovich, 1945). Another unique feature of the anopluran head is the **obturaculum,** an elastic, cupulate plug that separates the head cavity from the thorax. Retractile muscles originating on the obturaculum are inserted on the proximal arms of the piercing stylets (Stojanovich, 1945).

Piercing-sucking mouthparts, dipteran types

Mouthparts of flies are basically suctorial in nature, with both piercing and nonpiercing types represented. There is generally a wide diversity of mouthpart structures among Diptera. Internally, the preoral area is developed into a strong sucking diaphragm pump, the **cibarial pump.** Externally, the elongated piercing stylets are sheathed by the conspicuous labium except where the labium itself is the principal piercing structure. In some families mouthparts are reduced, or vestigial, and nonfunctional in the adult flies (e.g., Gasterophilidae and many Oestridae).

Three subtypes characterize adult dipteran mouthparts: the horse fly–mosquito subtype, the house fly subtype, and the stable fly subtype. These subtypes are described below. Study fresh or slide-mounted specimens representing each subtype.

Horse fly subtype (Figure 2.6)

This is the generalized subtype, which is both piercing and sponging in function and represents the mouth-parts of hematophagous lower Diptera (e.g., mosquitoes, black flies, punkies, sand flies, snipe flies, and horse flies). Visualize the basic elements of the chewing-type mouthparts as being elongated to form a proboscis; the finer elements are ensheathed in a prominent labium. When this type of proboscis is examined in cross section, the similarity of mouthparts arrangement to that of the chewing type is made clearer.

The **labrum** is an elongated passive structure that closes the roof of the food canal. Immediately behind the labrum are the **mandibles.** In most groups of blood-sucking lower Diptera the mandibles operate in a transverse plane much as do those of the basic chewing-type insect. However, their operation is more analogous to the shearlike snipping of scissors because of the contiguous interaction of the flattened mandibular blades. The scissoring movement permits the mandibles to slide between the apposed labrum and **hypopharynx.**

In mosquitoes (Figure 2.5) the mandibles move within a vertical, or probing, plane resulting from a change in position of mandibular musculature (Hudson, 1970). They move vertically adjacent to the labrum and apposed hypopharynx. In this case the mandibles do not appear to play a major role in probing entry, but serve to shield the maxillary teeth during withdrawal of the stylets.

The **hypopharynx** is a single median stylet traversed by the salivary duct and connected basally to the **salivary pump.** It generally forms the floor of the food canal by more-or-less close adherence to the exposed **epipharyngeal groove** of the labrum. In mosquitoes, however, the labrum and hypopharynx form a rather tight seal to produce the food canal.

The **maxilla** is developed from the maxillary lacinia of chewing-type mouthparts. The paired maxillae are the principal probing and anchoring structures of the piercing stylets. They lie ventrolaterad to the labrum and move alternately to pull the stylets into the feeding substrate through the anchoring and cutting action of their distal, posteriorly directed teeth (Figure 2.5).

The **labium** is the largest of the mouthparts and is homologous to the chewing-type prementum. It is unpaired, and the dorsal gutter receives the slender piercing stylets when in repose. During blood feeding the labium is elbowed back to expose the biting stylets. At the apex of the labium is a pair of flexible lobes, the **labella,** which represents the paraglossa of more generalized mouthparts. In the act of blood feeding, the labella serve to support the piercing stylets at the point of skin entry. In many flies the labellar surface is traversed by closely set, fine channels (**canaliculi**) directed toward the embraced stylets. Canaliculi direct surfacing blood to the food canal to be sucked up during feeding.

A pair of internal head pumps produces the suction needed to draw blood from the feeding substrate (Figure 2.6). The **cibarial pump,** the larger of these, contracts alternately with the **pharyngeal pump** to shuttle blood into the **esophagus.** Cibarial pump muscles originate on the clypeus and insert on the anterior wall of the pump. Both pumps are of a simple diaphragm-type structure with a rigid wall and a flexible wall. The flexible wall is drawn by muscle action to open the lumen of the pump, and relaxation causes the wall to close the lumen by inherent elasticity. A similarly constructed, third, internal head pump, the **salivary pump,** lies posterior to the rigid wall of the cibarial pump. Muscles to operate the salivary pump originate on the rigid wall of the cibarial pump. Action of the salivary pump forces secretions through the hypopharynx to be released at the tip of the stylets (Bonhag, 1951).

Make a parasaggital cut and dissection of the head of a freshly killed or preserved horse fly. The three pumps can be located and their action determined. Only females have well-developed mandibles in this type of mouthparts, and only females will pierce tissue to suck blood.

House fly subtype (Figure 2.7)

The majority of Diptera possess the nonpiercing, sponging type of mouthparts exemplified by the house fly and blow fly. In this subtype both males and females have similar oral structures with small differences. The large fleshy proboscis, formed by the **labium,** is most prominent. The labium terminates in a pair of sponging or rasping organs, the **labella** (like that seen in the horse fly subtype). Minute grooves in the labellar surface function like canaliculi of the horse fly labium, but are called **pseudotracheae.**

With the exception of the **maxillary palpi,** both mandibles and maxillae are inconspicuous, having been incorporated into basal elements of the labium and preoral area. Lying on top of the grooved labium is the short, spadelike **labrum-epipharynx.** The **hypopharynx** is pressed close posteriorly to the labrum and thus forms the **food canal** with the deeply grooved epipharynx.

Internally there are only two pumps: the enlarged, heavily muscled cibarial pump and a smaller salivary pump. Make a parasaggital section of the head of an adult blow fly or flesh fly and compare the internal structures with those of the horse fly subtype.

Stable fly subtype (Figure 2.8)

This subtype of dipteran mouthparts is found in the stable **fly**, horn fly, tsetse fly, and the ked, or louse fly, and its relatives. This subtype is specialized for piercing only and consists of a rigid proboscis formed from the **labium, labrum,** and **hypopharynx;** it has the same elements as the house fly subtype, but the labium is the principal piercing structure. The small **labellar lobes** at the tip of the labium are armed with rasping denticles, **prestomal teeth,** which cut into the substrate and anchor the proboscis during blood feeding. The food canal is formed by the ventral groove of the labrum-epipharynx in close apposition to the labial gutter. As with other subtypes, the hypopharynx is traversed in length by the salivary duct. **Maxillary palps** are located at the base of the proboscis, but maxillae and mandibles have been incorporated into basal elements of the labium.

In the louse fly (Figure 2.9) and related bat flies (Streblidae and Nycteribiidae), this rigid type of proboscis is retracted as a single unit into a deep internal pouch in the ventral head region and is protracted for feeding.

Piercing mouthparts, siphonapteran type (Figure 2.11)

The general appearance of the mouthparts of fleas belies the close phylogenetic relationship of the Siphonaptera to the Diptera. The labrum is rudimentary, and the piercing stylets are formed from the **maxillary laciniae** and the elongated **epipharynx.** Both sexes of adult fleas are obligate blood feeders, and both possess piercing-type mouthparts.

The epipharynx is an unpaired stylet rigidly fixed to the prestomal area. Its proximal loop with a flexible anterior wall forms the frame and diaphragm for the cibarial pump (Snodgrass, 1946). The floor of this pump is closed by the short **hypopharynx.** The epipharynx is blunt apically and is grooved behind to form the food canal with the closely appressed concaved laciniae. In addition to the bladelike toothed laciniae, the maxillae include a pair of large, palpus-bearing **lateral lobes** (wrongly termed "mandibles" by some). Mandibles are absent in the adult flea. Salivary secretions released from the tip of the short hypopharynx are conducted along the stylets by a pair of microscopic lacinial grooves.

The simple labium with its conspicuous terminal, segmented palpi serves as a protective sheath for the stylets in repose. In the act of feeding, the labium is bent forward, and its palpi are elbowed back so that the penetrating stylets are cradled in the notch of the labial prementum (Figure 2.11). As in the horse fly subtype of dipteran sucking apparatus, a second internal pump, the pharyngeal pump, is developed posterior to the true mouth at the proximal end of the cibarial pump.

Piercing-sucking mouthparts, lepidopteran type (Figure 2.10)

There are very few known skin-piercing or blood-sucking species of the large order Lepidoptera. That some moths may suck blood or habitually pierce the skin of mammal hosts was reported in convincing detail by Banziger (1968, 1970). The piercing mouthparts of the blood-sucking species have been modified from those of a larger group of fruit-piercing moths. Unlike other mouthpart types described, the fascicle in these moths is formed from only one pair of basic oral elements, the maxillae. The piercing stylet, a long flexible proboscis coiled in repose, consists of the opposed concaved surfaces of the hollow maxillary galeae. The appressed galeae form the food canal, but there is no independent salivary duct as in other mouthpart types. These moths lack a hypopharynx.

Extension of the proboscis (paired galeae) into a protracted feeding and probing attitude (Figure 2.10) is accomplished largely by internal fluid pressure in each hollow galea body. Penetration is made by rapid spindlelike flexing of the extended tip of the proboscis pressed against the feeding substrate. This produces superficial tearing at the entry site by the apical hooks of the paired galeae. Probing is made by combined rotating action of the head and basal maxillary sclerites so that the galeae slide against each other alternately. They are pulled deeper into tissue by anchoring action of their erectile lateral barbs. This is similar to lacinial penetration in the dipteran and siphonapteran types. During feeding the moth draws the entire proboscis up and down in the wound to produce continual blood flow with the galeal rasping spines. Salivary secretions are introduced through the food canal or by free flow down the outside of the proboscis (Banziger, 1970).

Piercing-sucking mouthparts, Acari, Ixodida type (Figure 2.12)

Although mites and ticks are not insects, they are of great importance to medical entomology and possess basic mouthparts that cannot be homologized with those of insects. Acarines, like other arachnids, imbibe only fluids.

Solids must be liquefied externally by salivary secretion on the substrate, or in the preoral cavity, for ingestion. The mouthpart-bearing section of the acarine is known as the **gnathosoma** or **capitulum.** Essentially the capitulum consists of a sclerotized ringlike collar bearing a pair of segmented palpi (or pedipalps) laterally and a medial pair of protractible toothed, cylindrical shafts (chelicerae). The basis capituli is composed largely of the fused coxae of the palpi. The preoral cavity is closed dorsally by the **epistome** and ventrally by the **hypostome.** In ticks the hypostome is enlarged and elongated with prominent teeth ventrally that anchor the mouthparts in the feeding substrate.

According to Gregson (1960) the process of feeding by *Dermacentor andersoni* may be subdivided into three periods: attaching, feeding, and detaching.

Attaching involves the secretion of a milky fluid that hardens quickly into a latex-like material forming a papilla that molds the exterior oral structures to the feeding substrate. Feeding is done by intermittent probing and cutting of subdermal tissues by the digits. Clear salivary fluid is introduced into the wound during this stage and acts as an anticoagulant or changes capillary permeability. Detaching results when the chelicerae are withdrawn, thus decreasing pressure by other mouthparts against the molded papilla and allowing the tick to back away.

Salivary secretions are introduced directly into the preoral cavity or through a pair of long slender **salivary stylets** located between the palpi and hypostome. The chelicerae are terminated by one movable and one immovable toothed digit (**chelae**) and are protracted to tear and probe by internal fluid pressure. Retraction of the chelicerae is by muscle action.

The adult ixodid tick can be examined as a large, easily obtained example of the generalized parasitiform acarine mouthparts.

Maggot mouthparts, larvae of cyclorrhaphous Diptera (Figure 2.13)

In general the mouthparts of those immature insects with gradual metamorphosis are smaller replicas of adult chewing or piercing mouthparts. Immature insects having complete metamorphosis generally have chewing-type mouthparts regardless of the type found in adults. However, maggots, the larvae of cyclorrhaphous or higher flies, possess peculiar, more modified oral structures – the cephalopharyngeal skeleton and mouth-hooks. These oral structures are so changed from a basic chewing function that their homologies are difficult to recognize. Because many maggots also are agents of myiasis, their uniquely adapted oral structures warrant closer examination.

Larvae of cyclorrhaphous Diptera typically have only a paired set of **mouth-hooks** as externally visible oral structures. These mouth-hooks articulate internally with the H-shaped **labial-hypopharyngeal sclerite** which embraces a pair of median **malar sclerites** (cardo) anteriorly and joins with the ventroanterior arms of the **paraclypeal phragmata** posteriorly. The phragmata, with its rigid ventral wall and strongly muscled and flexible dorsal wall, functions as a diaphragm pump much as does the cibarial pump of adult Diptera (Figure 2.6). The salivary duct opens close to the labial-hypopharyngeal sclerite. Scratching, probing, and tearing by the mouth-hooks (formed mostly from the maxillae) moves food particles in a fluid medium of salivary secretions into the **oral atrium.** Sucking action by the cibarial pump draws the liquid diet into the anterior digestive tract (Menees, 1962). Mouth-hooks also function for anchoring the maggot in soft substrate or as grappling hooks for creeping locomotion. Examine the larval cephalopharyngeal structures and mouth-hooks of representative higher Diptera (e.g., late-stage larvae of Calliphoridae, Sarcophagidae, Muscidae, Gasterophilidae, or Cuterebridae).

References

Banziger, H. 1968. Preliminary observations on a skin-piercing blood-sucking moth (*Calyptra eustrigata* (Hmps.)(Lep., Noctuidae)) in Malaya. *Bull. Entomol. Res.* 58:159–63, plates 9, 10.

- 1970. The piercing mechanism of the fruit-piercing moth *Calpe* (*Calyptra*) *thalictri* Bkh. (Noctuidae) with reference to the skin-piercing blood-sucking moth *C. eustrigata* Hmps. *Acta Trop.* 27:54–88.

Bonhag, P. F. 1951. The skeleto-muscular mechanism of the head and abdomen of the adult horsefly (Diptera-Tabanidae). *Trans. Am. Entomol. Soc.* 77:131–202.

Buckup, L. 1959. Der Kopf von *Myrsidea cornicis* (deGeer) Mallophaga-Amblycera. *Zool. Jb. Anat.* 77:241–88.

Dickerson, G., and M. M. J. Lavoipierre. 1959a. Studies on the methods of feeding of blood-sucking arthropods. II. The method of feeding adopted by the bedbug (*Cimex lectularius*) when obtaining a blood meal from the mammalian host. *Ann. Trop. Med. Parasitol.* 53:347–57.

- 1959b. Studies on the methods of feeding of blood-sucking arthropods. III. The method by which *Haematopota pluvialis* (Diptera, Tabanidae) obtains its blood meal from the mammalian host. *Ann. Trop. Med. Parasitol.* 53:465–72.

Evans, G. O., J. G. Sheals, and D. MacFarlane. 1961. *The Terrestrial Acari of the British Isles*, vol. I. British Museum (Natural History), London.

Evans, G. O., and W. M. Till. 1965. Studies on the British Dermanyssidae (Acari: Mesostigmata). I. External morphology. *Bull. Brit. Mus.* (*Nat. Hist.*) *Zool.* 13(8):249–94.

Ferris, G. F. 1951. *The Sucking Lice*. Memoire, Pacific Coast Entomological Society, vol. I. California Academy of Science, San Francisco.

Fox, R. M., and J. W. Fox, 1964. *Introduction to Comparative Entomology*. Reinhold, New York.

Gregson, J. D. 1960. Morphology and functioning of the mouthparts of *Dermacentor andersoni* stiles. *Acta Trop.* 17:48–79.

Harwood, R. F., and M. T. James. 1979. *Entomology in Human and Animal Health*, 7th ed. Macmillan, New York.

Hudson, A. 1970. Notes on the piercing mouthparts of three species of mosquitoes (Diptera: Culicidae) viewed with the scanning electron microscope. *Canad. Entomol.* 102:501–12.

Lavoipierre, M. M. J. 1967. Feeding mechanism of *Haematopinus suis*, on the transilluminated mouse ear. *Exp. Parasitol.* 20:303–11.

Lavoipierre, M. M. J., G. Dickerson, and R. M. Gordon. 1959. Studies on the methods of feeding of blood-sucking arthropods. I. The manner in which triatomine bugs obtain their blood meal from the mammalian host. *Ann. Trop. Med. Parasitol.* 53:235–50.

Menees, J. H. 1962. The skeletal element of the gnathocephalon and its appendages in the larvae of higher Diptera. *Ann. Entomol. Soc. Am.* 55:607–16.

Snodgrass, R. E. 1935. *Principles of Insect Morphology*. McGraw-Hill, New York.

- 1944. *The Feeding Apparatus of Biting and Sucking Insects Affecting Man and Animals*. Smithsonian Misc. Collection, 104 (7).

- 1946. *The Skeletal Anatomy of Fleas* (*Siphonaptera*). Smithsonian Misc. Collection, 104 (18).

- 1948. *The Feeding Organs of Arachnida Including Mites and Ticks*. Smithsonian Misc. Collection, 110 (10).

Stojanovich, C. J., Jr. 1945. Head and mouthparts of the sucking lice. *Microentomology* 10:1–46.

Usinger, R. L. 1966. *Monograph of Cimicidae*. Thomas Say Foundation, Entomological Society of America, College Park, Md.

Wenk, P. 1953. Der Kopf von *Ctenocephalus canis* (Curt.). *Zool. Jb. Anat.* 73:103–64.

- 1962. Anatomie des Kopfes von *Wilhelmia equina*. L. (Simuliidae syn. Melusinidae, Diptera). *Zool. Jb. Anat.* 80:81–134.

2.1 *BASIC CHEWING-TYPE MOUTHPARTS*

<u>*Blattella germanica*</u> (Blattellidae)

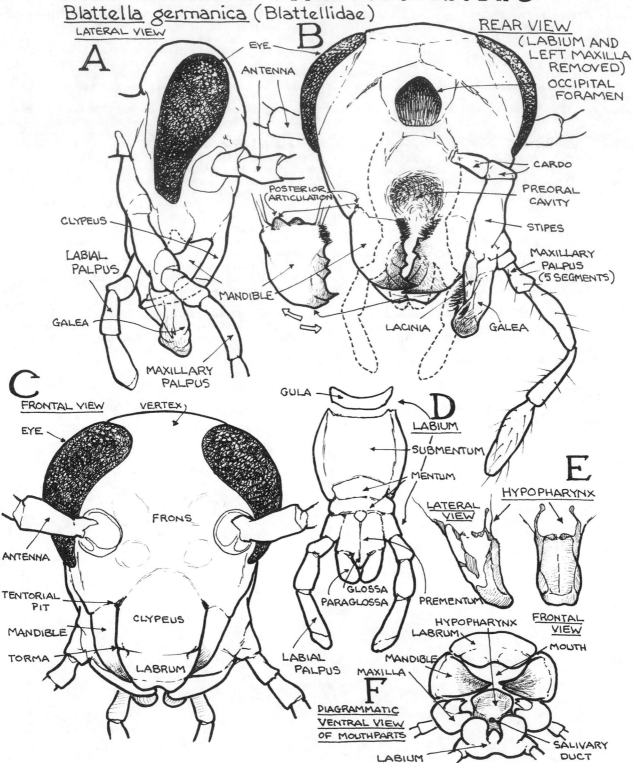

A <u>LATERAL VIEW</u>

EYE

ANTENNA

B <u>REAR VIEW</u> (LABIUM AND LEFT MAXILLA REMOVED)

OCCIPITAL FORAMEN

POSTERIOR (ARTICULATION)

CLYPEUS

LABIAL PALPUS

GALEA

MANDIBLE

MAXILLARY PALPUS

CARDO

PREORAL CAVITY

STIPES

MAXILLARY PALPUS (5 SEGMENTS)

LACINIA

GALEA

C <u>FRONTAL VIEW</u>

VERTEX

EYE

ANTENNA

FRONS

TENTORIAL PIT

MANDIBLE

TORMA

CLYPEUS

LABRUM

GULA

D <u>LABIUM</u>

SUBMENTUM

MENTUM

<u>LATERAL VIEW</u>

GLOSSA

PARAGLOSSA

PREMENTUM

LABIAL PALPUS

E HYPOPHARYNX

<u>FRONTAL VIEW</u>

MOUTH

HYPOPHARYNX

LABRUM

MANDIBLE

MAXILLA

F <u>DIAGRAMMATIC VENTRAL VIEW OF MOUTHPARTS</u>

LABIUM

SALIVARY DUCT

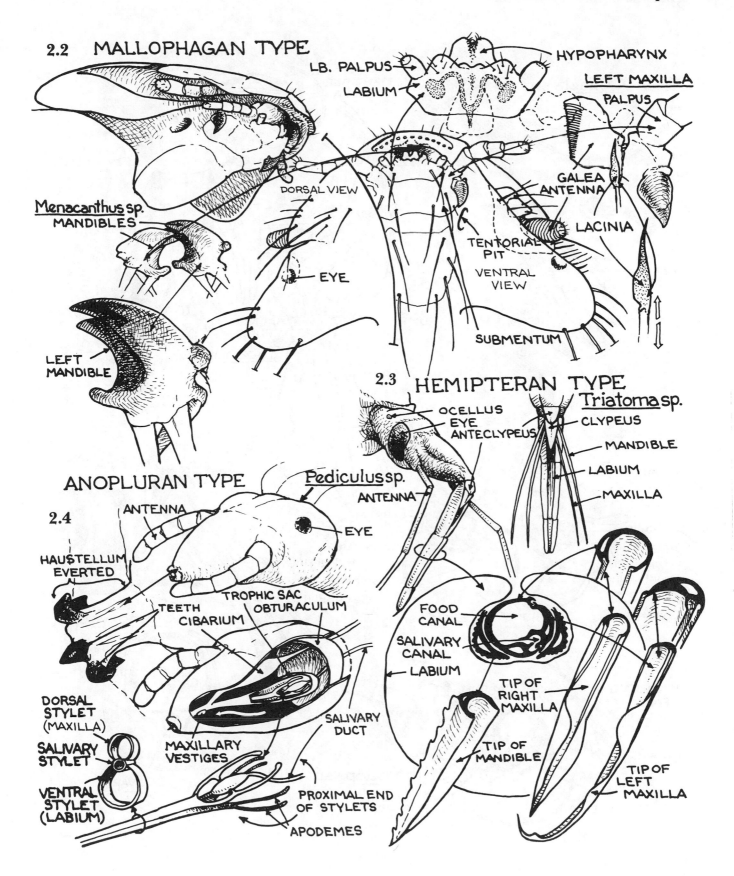

2.2 MALLOPHAGAN TYPE

LB. PALPUS
LABIUM
HYPOPHARYNX
LEFT MAXILLA
PALPUS

Menacanthus sp. MANDIBLES

DORSAL VIEW

GALEA
ANTENNA

EYE

LACINIA

TENTORIAL PIT

VENTRAL VIEW

LEFT MANDIBLE

SUBMENTUM

2.3 HEMIPTERAN TYPE *Triatoma* sp.

OCELLUS
EYE
ANTECLYPEUS

CLYPEUS
MANDIBLE
LABIUM
MAXILLA

ANOPLURAN TYPE

Pediculus sp.
ANTENNA

2.4

ANTENNA

EYE

HAUSTELLUM EVERTED

TROPHIC SAC
TEETH
CIBARIUM
OBTURACULUM

FOOD CANAL
SALIVARY CANAL
LABIUM

DORSAL STYLET (MAXILLA)

SALIVARY STYLET

VENTRAL STYLET (LABIUM)

MAXILLARY VESTIGES

SALIVARY DUCT

TIP OF RIGHT MAXILLA

TIP OF MANDIBLE

TIP OF LEFT MAXILLA

PROXIMAL END OF STYLETS

APODEMES

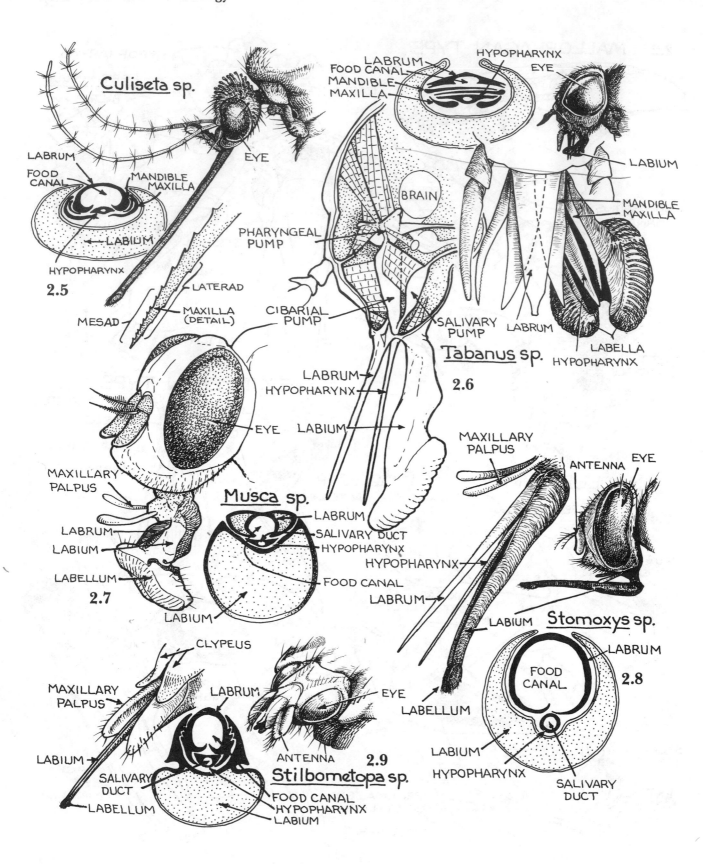

Culiseta sp.

LABRUM
FOOD CANAL
MANDIBLE
MAXILLA
EYE
LABIUM
HYPOPHARYNX

2.5

LATERAD
MESAD
MAXILLA (DETAIL)

LABRUM
FOOD CANAL
MANDIBLE
MAXILLA
HYPOPHARYNX
EYE

BRAIN

PHARYNGEAL PUMP

CIBARIAL PUMP

SALIVARY PUMP

LABRUM
HYPOPHARYNX
LABIUM

LABIUM

MANDIBLE
MAXILLA

LABRUM
LABELLA
HYPOPHARYNX

Tabanus sp.

2.6

MAXILLARY PALPUS

LABRUM
LABIUM
LABELLUM

2.7

EYE

Musca sp.

LABRUM
SALIVARY DUCT
HYPOPHARYNX
FOOD CANAL

LABIUM

MAXILLARY PALPUS

ANTENNA
EYE

HYPOPHARYNX

LABRUM

LABIUM

LABELLUM

Stomoxys sp.

LABRUM

2.8

FOOD CANAL

LABIUM
HYPOPHARYNX
SALIVARY DUCT

CLYPEUS

MAXILLARY PALPUS

LABRUM

LABIUM

SALIVARY DUCT
LABELLUM

EYE

ANTENNA
2.9
Stilbometopa sp.

FOOD CANAL
HYPOPHARYNX
LABIUM

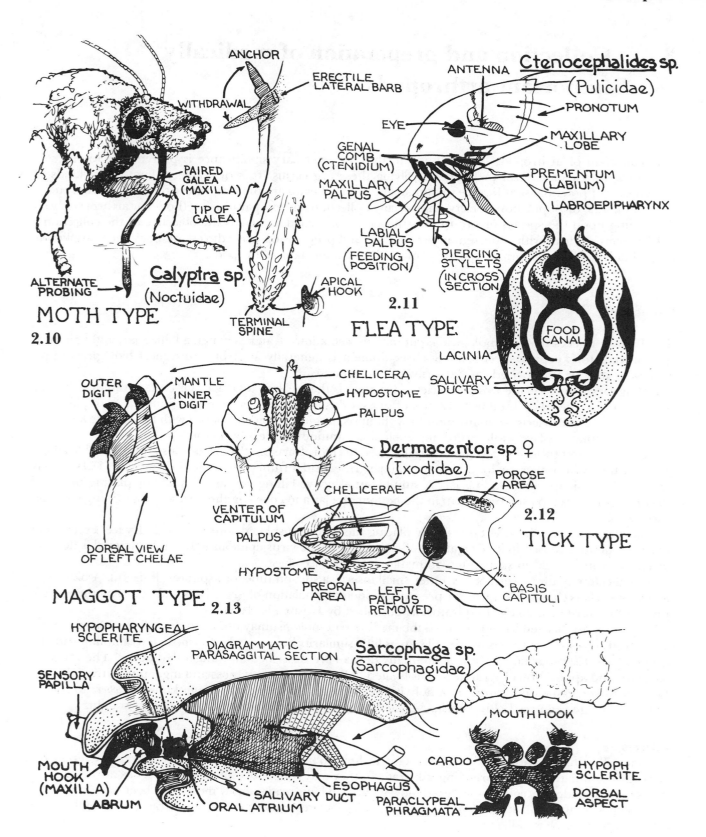

ANCHOR

ERECTILE
LATERAL BARB

WITHDRAWAL

PAIRED
GALEA
(MAXILLA)

TIP OF
GALEA

ALTERNATE
PROBING

Calyptra sp.
(Noctuidae)

APICAL
HOOK

TERMINAL
SPINE

MOTH TYPE

2.10

ANTENNA

Ctenocephalides sp.
(Pulicidae)

PRONOTUM

EYE

MAXILLARY
LOBE

GENAL
COMB
(CTENIDIUM)

PREMENTUM
(LABIUM)

MAXILLARY
PALPUS

LABROEPIPHARYNX

LABIAL
PALPUS
(FEEDING
POSITION)

PIERCING
STYLETS
(IN CROSS
SECTION)

2.11

FLEA TYPE

FOOD
CANAL

LACINIA

SALIVARY
DUCTS

OUTER
DIGIT

MANTLE

INNER
DIGIT

CHELICERA

HYPOSTOME

PALPUS

Dermacentor sp. ♀
(Ixodidae)

POROSE
AREA

CHELICERAE

VENTER OF
CAPITULUM

2.12

TICK TYPE

PALPUS

HYPOSTOME

DORSAL VIEW
OF LEFT CHELAE

PREORAL
AREA

LEFT
PALPUS
REMOVED

BASIS
CAPITULI

MAGGOT TYPE 2.13

HYPOPHARYNGEAL
SCLERITE

DIAGRAMMATIC
PARASAGGITAL SECTION

Sarcophaga sp.
(Sarcophagidae)

SENSORY
PAPILLA

MOUTH HOOK

MOUTH
HOOK
(MAXILLA)

CARDO

HYPOPH
SCLERITE

LABRUM

ORAL ATRIUM

SALIVARY DUCT

ESOPHAGUS

PARACLYPEAL
PHRAGMATA

DORSAL
ASPECT

3   Collection and preparation of medically important arthropods

The collection of arthropods of presumed medical or veterinary significance is prerequisite to any subsequent research or control programs involving these organisms. It is desirable to know what kinds of arthropods of potential significance to human and animal health are present, where they are located, and in what numbers, or population density, they occur. Collection and preparation methods for arthropods vary depending upon the species or group of arthropods concerned as well as the objective of the collection.

This exercise deals with representative collection and preparation procedures and equipment applicable to qualitative and/or quantitative surveys of medically important arthropods.

Collection

The basic but necessary entomological equipment is well known – an insect net, a killing jar, and a pinning box or vials for keeping specimens. Such equipment is generally available from most biological supply houses or may be improvised satisfactorily at little expense.

Killing jars may be made using a variety of materials that release toxic gas. Granular cyanide salt (e.g., sodium or potassium) covered with 12 mm of dry sawdust over which is poured a 6-mm layer of plaster of Paris slurry will kill insects for many months without recharging. Avoid all contact with the cyanide and work in a well-ventilated place while preparing such jars. Cyanide is a deadly poison.

Jars or large test tubes containing material soaked in chloroform or ethyl acetate also make suitable killing jars, but they will not last as long as a cyanide jar. Insecticidal fumigants such as dichlorvos (DDVP) may provide the toxic gas for a killing jar. A simple but effective killing jar can be made by placing pieces of DDVP-impregnated plastic resin strip in the bottom of a jar and covering them with a tight-fitting blotting-paper disk.

Specimens also may be killed quickly by placing them in an insulated container with dry ice. Freezing of adult mosquitoes is used in investigations of arthropod-borne virus epidemics. If viable material is desired, specimens can be held in suspended animation inside an ice chest.

For collecting adult mosquitoes or other small insects, a suction tube, or aspirator, is useful. A blow-type aspirator such as shown in Figure 3.1 prevents accidental inhalation of organisms and dust. Another type is a battery-powered mechanical aspirator, as described by Husbands (1958).

In addition to this general equipment, the medical entomologist may use a number of specialized devices or methods in a specific type of survey. There is little similarity between a kit for mosquito sampling and one used to collect fleas or lice. A few of the many devices are described and illustrated below. The range of variation and application of such devices is limited only by the circumstances and ingenuity of the collector. The key to successful collecting is not only in using effective devices, but also in knowing *where* to collect. This knowledge requires field experience and familiarity with the biology of the quarry sought.

Attractants

Some collecting devices employ an attractant, or lure, to induce arthropods to enter a trap. Generally the attractants simulate a host or breeding substrate, or they may consist of a pheromone, or sex attractant. Dry ice, carbon dioxide gas, color, heat, light, movement, and baits made of raw meat, stale beer, or rotting fruit are all types of attractants. Some methods use a living but restrained host as the attractant in combination with a trap or an alert collector.

Trapping flying arthropods

The **Malaise trap** (Figure 3.2) is a widely used device for sampling all types of flying insects, especially Diptera. It may be used alone or baited with attractants such as carbon dioxide. A variation of this trap is the **Manning trap** (Figure 3.3), designed for tabanid sampling. A shiny black spherical or cylindrical target serves as the attractant. Dry ice is the attractant used in both the **DeFoliart–Morris trap** (Figure 3.4) and the **canopy trap** (Figure 3.5), although other attractants may be employed. There are several designs for mounted **boat or car traps** to collect flying insects while such vehicles are moving from 30 to 42 km per hour (Sommerman and Simmet, 1965).

The **Bishopp fly trap** (Figure 3.6) consists of a screen cone under a vertical cage and is baited with carrion or feces. The **lard can trap** (Figure 3.7) is a horizontal cylinder with screen entry cones at each open end (Bellamy and Reeves, 1952). A bait or lure (e.g., dry ice, a live host, or carrion) is placed on the floor of the cylinder. The **New Jersey light trap** (Figure 3.8) is a vertical cone and cylinder device using light as the lure and a small fan to force insects into the cone. This trap is a standard sampling device for adult mosquitoes and is copied in principle by the battery-operated, collapsible **CDC portable light trap** (Figure 3.9). Dry ice may be used with both of these light traps as a supplemental lure. The U.S. Army has developed a solid state mosquito light trap (Driggers et al., 1980), which is the principal element of the U.S. Army Portable Insect Survey Set (Kardatzke et al., 1980). Another innovation to the portable light trap is one that utilizes a chemical source of light, thus saving on energy needs for trap operation (Service and Highton, 1980).

The **ramp trap** (Figure 3.11) is designed for directional sampling of flying insects without the need for an attractant. The **pipe trap** (Nelson, 1980) uses a short section of asbestos-cement pipe and a wood and foam rubber piston-plunger to collect resting adult mosquitoes from animal burrows (Figure 3.15). **Emergence traps** (Figure 3.10) will capture flying insects as they leave their pupal habitat as an imago. Their effectiveness depends on the collector's knowing where the immatures develop.

A more elaborate type of **immigration and emergence trap,** such as is shown in Figure 3.17, is designed to sample arthropods attracted to and developing in carrion or dung with minimal disruption of arthropod succession. The trap provides ingress and egress funnel portals. The egress portals and half of the ingress portals collect arthropods in a pitfall-like device containing antifreeze solution (Schoenly, 1980).

Collecting crawling arthropods

The **tick drag** (Figure 3.12) is used to collect unengorged ticks and is made of a square meter of flannel-like material attached on one side to a 1.2-m drawbar. Unattached chiggers may be collected with the chigger panel, a square meter of opaque material having a 7.6-cm-square center window. When the panel is laid on the ground, the chiggers will crawl to the underside of the window. The **carbon dioxide tick trap** of Hokama and Howarth is effective in collecting certain soft ticks such as *Ornithodoros coriaceus*. It consists of a rectangular enamel pan with straight sides; in the center of the pan is placed a rectangular plastic box with perforated sides. The pan is set into the ground of the collecting area so that the adjacent soil is flush with the top rim of the pan. A piece of dry ice placed in the plastic box serves as the source of CO_2 gas (Figure 3.14).

A modified **CO_2 burrow trap** for collecting ticks or fleas from animal burrows is shown in Figure 3.18.

Arthropods, such as wolf spiders, ticks, beetles, and cockroaches, that run or crawl over the surface of the ground can be collected in any of various types of pitfalls. A simple **pitfall trap** is shown in Figure 3.16 (Morrill, 1975).

The principle of the **Berlese funnel** (also called the **Tulgren funnel**) is used in many collecting devices (Figure 3.13). The device employs the repellent properties of light, heat, desiccation, or a fumigant to drive living organisms from the material in which they are hidden, down the funnel, and into a collecting jar. This may be used to separate mites from duff and debris, fleas from bedding, bugs from bird nests, or larvae from sod.

Collecting mosquito immatures

An old-fashioned long-handled dipper is standard equipment for collecting most mosquito larvae. When the dipper is pushed under the water surface at an angle (approximately 45°), water flowing into the dipper basin carries mosquito "wigglers" with it. A large-tipped "eye-dropper" pipette or urethral syringe can be used to collect larvae from inaccessible places, such as tree holes, and also is useful for transferring larvae from the dipper to sample vials.

Larvae and pupae of *Coquillettidia* and *Mansonia* attach to submerged parts of aquatic plants and can be collected by a washing technique described by Hagmann (1952).

Mosquito eggs that are laid as "rafts" or "boats" may be lifted from habitat water, transported to the laboratory on wet blotting paper, and reared in water-filled pans. Eggs that are laid singly can be collected by a washing-flotation technique. The **Horsfall egg separator** is a machine designed specifically for this purpose. It uses a revolving wire-cloth wash drum, a series of sieves, and a sugar or salt solution to float eggs free from other sediment after concentration.

Several species of adult mosquitoes can be collected with the **resting box.** The box, 0.3 m³ in size, generally is made of wood and has one open side (Burbutis and Jobbins, 1958). Mosquitoes will enter the box to rest during the morning and may be trapped inside and chloroformed. Such resting boxes, along with light traps, are standard items of equipment for most mosquito surveys, especially when blood-engorged specimens are desired.

Collecting ectoparasites from small mammals and birds

The collecting of ectoparasites from living small mammals or birds can be facilitated by hand-holding the host over a white enamel collecting tray and brushing its pelage vigorously with a "nit-comb" or a stiff bristled brush (e.g., an old toothbrush is quite suitable). A few puffs of pyrethrin dust (0.5 – 1.0%) directed into fur or feathers will facilitate the removal of most ectoparasites. A small pyrethrin puff-duster is also useful for similar collecting from live animals that cannot be handled easily after trapping. Such hosts may be left in the trap, dusted with pyrethrin, and brushed while the collector holds the traps containing the host over a collecting tray. Another method of collecting from small living hosts is to insert all but the head of the host into a polyethylene bag containing a cotton pledget soaked in chloroform or ethyl acetate. Struggling by the restrained host will help in the dropping-free of most moribund or dead ectoparasites in a few minutes. The bag may be slit for closer examination after the host is removed and all ectoparasites transferred to vials of 70% ethyl alcohol by means of a small moistened brush or forceps. A suggested list of equipment for ectoparasite collecting is given below.

A standardized technique for removal of ectoparasites from small dead mammals or birds employs a detergent wash of the entire specimen. The dead host is placed in a closed container with a cotton ball to which a few drops of chloroform are added. After a few moments the animal is transferred to a cylinder of water containing a few drops of liquid detergent. The jar is closed and shaken vigorously. On standing, the sediment and ectoparasites will settle to the bottom of the liquid, and the dead host is removed. The liquid is stirred again and filtered (No. 2 filter paper) or poured through a fine-mesh sieve. Alternatively, the supernatant fluid may be poured off and the sediment poured into Petri dishes for examination under the dissecting microscope.

Where chiggers or ticks remain attached to dead hosts, a snip of skin or an entire ear or leg can be removed and preserved in alcohol pending subsequent removal of specimens in the laboratory.

Another method for ectoparasite collecting is the complete-host-dissolving technique (Watson and Amerson, 1967). This permits total recovery of the ectoparasite fauna, but the host itself is completely destroyed as a study specimen. The dissolving solution consists of 500 ml distilled water, 15 g dibasic anhydrous sodium phosphate (NaHPO$_4$), and 6 g trypsin. The pelt or skin and feathers of the host are removed and placed in the dissolving solution at 38°C for at least 12 hours. Potassium hydroxide (15 g) is added, and the "soup" is boiled until all large solid material is dissolved. The mixture is cooled, poured through a 22-mesh screen, and the residue examined. Caution is necessary with this procedure as the dissolving solution fumes are toxic.

Equipment Used for Collecting Ectoparasites in the Field

Paper or plastic bags	Scissors
Cotton and chloroform or ethyl acetate	Vials
White pan or tray	Labels, pencil, India ink, pen
Small plastic jar	Berlese funnel
Hand lens	Alcohol (ethyl or isopropyl)
Plastic funnel	Glycerin
Comb	Puff duster
No. 2 filter paper	Pyrethrin powder
Stiff brush or fine probe	Fine-tipped paint brush (sizes 00–1)
Fine-tipped forceps	Household detergent or soap
Syringe or pipette	

Modified from Watson and Amerson (1967).

Storage and preparation

Mites, ticks, fleas, lice, cockroaches, spiders, scorpions, and soft-bodied larvae of insects are among those forms best preserved in 70% ethyl or isopropyl alcohol. Heavily sclerotized or blood-engorged specimens may be "cleared" prior to study by placing them in aqueous 10% potassium hydroxide for 4–24 hours at room temperature. If the solution is warmed, a shorter period will suffice. After clearing, specimens are gently washed in water and transferred to acidified 70% alcohol.

Specimens may be mounted on microscope slides directly from water or 70% alcohol when **Hoyers** medium is used. Before being mounted in a medium of **Canada balsam,** specimens should be dehydrated for 15 minutes in each of the following: 70%, 85%, 95%, and absolute alcohol. They are then immersed in a clearing agent, such as cellusolve, beechwood creosote, eugenol, or xylol, before being mounted in balsam. Hardening of mountant on freshly prepared permanent slides may be hastened by warming the slides (40°C) for a week or more. Temporary microscope slide mounts can be made using water or glycerin as the mountant. All permanent slides should be labeled as shown in Figure 3.21.

Freeze drying of soft-bodied specimens is a method of dry-preserving a specimen with optimal retention of form and color. It is based on the sublimation of frozen fluids. The process involves the complete dehydration of frozen specimens in a vacuum (Hower, 1967).

Medium to large hard-bodied insects are preserved by impaling them on standard insect pins and letting them air dry. Freshly killed specimens may be pinned directly, but those that have dried for a couple of days or more after killing should be "relaxed" before pinning. A simple **relaxing chamber** can be made by placing wet sand or tissue paper in the bottom of a large jar, adding a few drops of carbolic acid to inhibit mold, and using a wire screen or handful of glass beads or marbles to support specimens above the moist substrate. Specimens to be relaxed are placed in an open container on the screen or the marbles, and the jar is closed tightly. After a few days specimens will be soft enough for pinning.

Larger adult Diptera and Hymenoptera should be pinned, as shown in Figure 3.23, by inserting the pin through the thorax slightly to the right of the midline. Coleoptera are pinned through the base of the right elytron. Small specimens, such as mosquitoes or black flies, are mounted on small pins known as minuten nadeln (Figure 3.19). These are best used only on freshly killed specimens. Minute Diptera (e.g., chloropids, ceratopogonids, and streblids) are best kept unpinned in vials of alcohol. For relaxed or brittle older specimens, use card points (Figure 3.20) or direct gluing to a standard insect pin (Figure 3.22). A useful adhesive for this purpose is clear fingernail polish. Proper and complete labeling is as important as proper mounting. Essential data such as the location and date of capture, the collector's name, and the host animal involved (if appropriate) are necessary for further study of any specimen (Figure 3.23).

Pinned specimens can be stored in standard insect boxes (Schmitt boxes) or in museum cases. Specimen boxes should be nearly airtight and protected from infestation by insects such as dermestid beetles. Naphthalene flakes or paradichlorobenzene (PDB) crystals in a perforated or fabric container fastened inside of the box will prevent other insects from feeding on stored specimens. Alternatively, a mothball (naphthalene) impaled on the head of a heated pin or a 25-mm square of dichlorvos-impregnated plastic (DDVP) on a pin also will repel dermestid beetles. The effectiveness of DDVP squares lasts for 1 year without their replacement.

References

Belkin, J. N., C. S. Hogue, P. Galindo, T. H. G. Aitken, R. X. Schick, and W. A. Powder. 1965. Mosquito studies (Diptera, Culicidae). II. Methods for the collection, rearing and preservation of mosquitoes. *Contrib. Am. Entomol. Inst. 1*(2): 19–78.

Bellamy, R. E., and W. C. Reeves. 1952. A portable mosquito bait trap. *Mosquito News 12*:256–8.

Bidlingmayer, W. L. 1974. The influence of environmental factors and physiological stage on flight patterns of mosquitoes taken in the vehicle aspirator and truck, suction, bait and New Jersey light traps. *J. Med. Entomol. 11*:119–46.

Borror, D. J., D. M. DeLong, and C. A. Triplehorn. 1976. *An Introduction to the Study of Insects,* 4th ed. Holt, Rinehart & Winston, New York.

Bram, R. A. 1978. *Surveillance and Collection of Arthropods of Veterinary Importance.* Agric. Handbook No. 518. U.S. Dept. of Agriculture, Washington, D.C.

Burbutis, P. P., and D. M. Jobbins. 1958. Studies on the use of a diurnal resting box for the collection of *Culiseta melanura* (Coq.). *Bull. Brooklyn Entomol. Soc. 53*:53–8.

Crockett, D. B. 1967. A method for collecting feather lice (Mallophaga). *J. Kansas Entomol. Soc. 40*:192–4.

DeFoliart, G. R., and C. D. Morris. 1967. A dry ice baited trap for the collection and field storage of haematophagous Diptera. *J. Med. Entomol. 4*:360–2.

Driggers, D. P., R. J. O'Connor, J. T. Kardatzke, J. L. Stup, and B. P. Schiefer. 1980. The U.S. Army solid state mosquito light trap. *Mosquito News 40*:172–8.

Gillies, M. T. 1969. The ramp-trap, an unbaited device for flight studies of mosquitoes. *Mosquito News 29*:189–93.

Hagmann, L. E. 1952. *Mansonia perturbans,* recent studies in New Jersey. *Proc. N.J. Mosq. Extermination Assoc. 39*:60–5.

Hokama, Y., and J. A. Howarth. 1977. Dry ice (CO_2) trap for efficient field collection of *Ornithodoros coriaceus* (Acarina: Argasidae). *J. Med. Entomol. 13*:627–8.

Hower, R. O. 1967. The freeze-dry preservation of biological specimens. *Proc. U.S. Natl. Mus. 119*(3549), 24 pp.

Husbands, R. C. 1958. An improved mechanical aspirator. *Calif. Vector Views 5*:72–3.

Kardatzke, J. T., D. P. Driggers, R. J. O'Connor, J. L. Stup, and B. P. Schiefer. 1980. The U.S. Army portable insect survey set. *Mosquito News 40*:178–80.

Lehker, G. E., and H. O. Deay. 1958. *Insects–How to Collect, Preserve and Identify Them.* Purdue Univ. Extension Bull. 352. Purdue University, Lafayette, Ind.

Miles, V. I. 1968. A carbon dioxide bait trap for collecting ticks and fleas from animal burrows. *J. Med. Entomol. 5*:491–5.

Moore, L. D., G. S. Staines, and W. M. Rogoff. 1977. The Cavanaugh trap for collection of eye gnats, *Hippelates collusor* (Townsend). *Calif. Vector Views 24*:1–5.

Morrill, W. L. 1975. Plastic pitfall trap. *Environ. Entomol. 4*:596.

Nelson, R. L. 1980. The pipe trap, an efficient method for sampling resting adult *Culex tarsalis* (Diptera: Culicidae). *J. Med. Entomol. 17*:348–51.

Oldroyd, H. 1958. *Collecting, Preserving and Studying Insects.* Hutchinson, London.

Peterson, A. 1934. *A Manual of Entomological Equipment and Methods,* part 1. Edwards Brothers, Ann Arbor, Mich. 48 pp. and 138 plates.

Schoenly, K. 1980. A demographic bait trap. Laboratory for Environmental Biology, University of Texas, El Paso, Tex. 8 pp.

Service, M. W., and R. B. Highton. 1980. A chemical light trap for mosquitoes and other biting insects. *J. Med. Entomol. 17*:183–5.

Smart, J. 1962. *Insects, Instructions for Collectors,* 4th ed. British Museum (Natural History), London.

Sommerman, K. M., and R. P. Simmet. 1965. Car-top insect trap with terminal cage in auto. *Mosquito News 25*:172–82.

Sudia, W. D., and R. W. Chamberlain. 1962. Battery operated light trap, an improved model. *Mosquito News 22*:126–9.

Sudia, W. D., R. D. Lord, and R. O. Hays, 1970. *Collection and Processing of Medically Important Arthropods for Arbovirus Isolation.* U.S. Dept. of Health, Education and Welfare, Communicable Disease Center, Atlanta.

Townes, H. 1972. A light-weight malaise trap. *Entomol. News 83*:239–47.

Watson, G. E., and A. B. Amerson, 1967. Instructions for collecting bird parasites. Smithsonian Inst. Information Leafl. 477. U.S. Museum of Natural History, Washington, D.C. 12 pp.

Yescott, R. E. 1977. A vacuum technique for estimating populations of Anoplura and Siphonaptera on California ground squirrels. *Calif. Vector Views 24*:19–22.

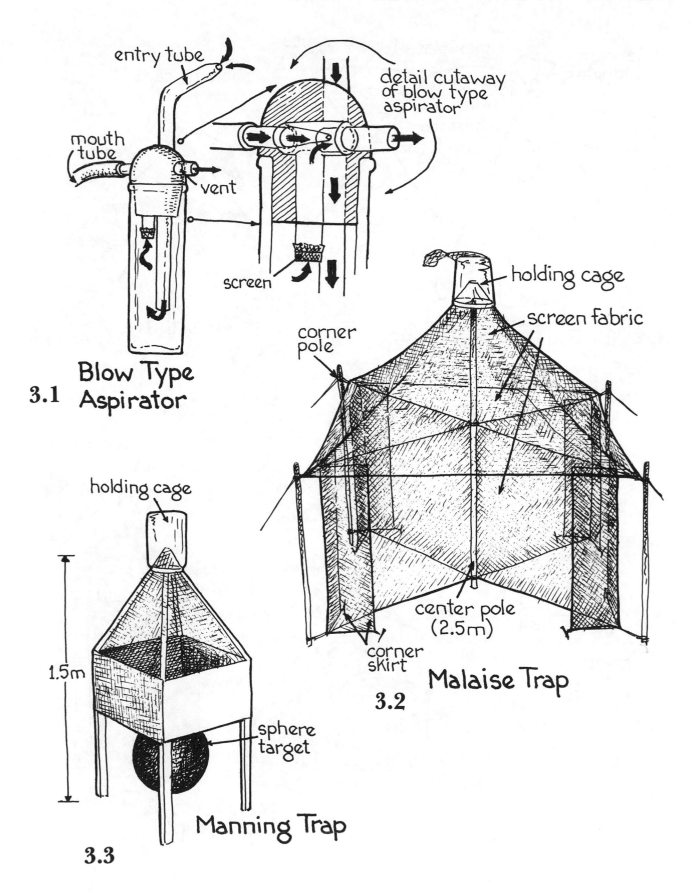

entry tube

mouth tube

vent

detail cutaway of blow type aspirator

screen

Blow Type
3.1   Aspirator

holding cage

screen fabric

corner pole

center pole
(2.5m)

corner skirt

Malaise Trap
3.2

holding cage

1.5m

sphere target

Manning Trap

3.3

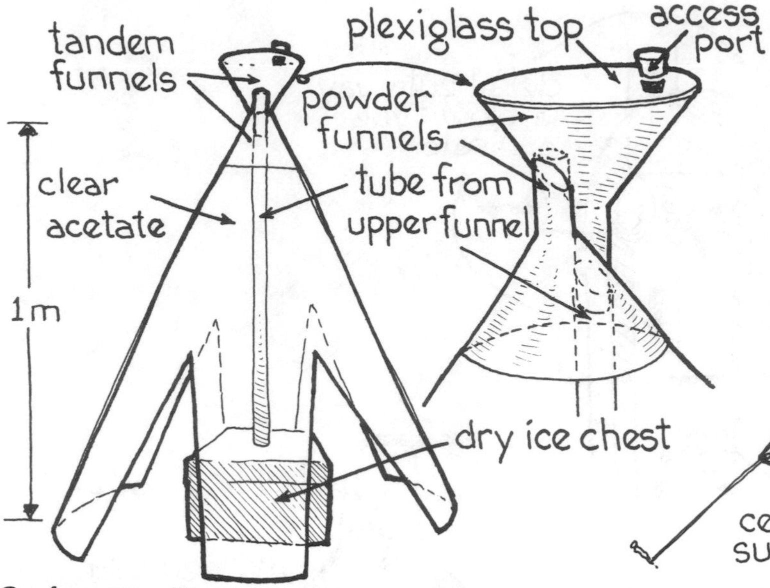

tandem funnels
plexiglass top
access port
powder funnels
clear acetate
tube from upper funnel
1 m
dry ice chest

3.4 DeFoliart-Morris Trap

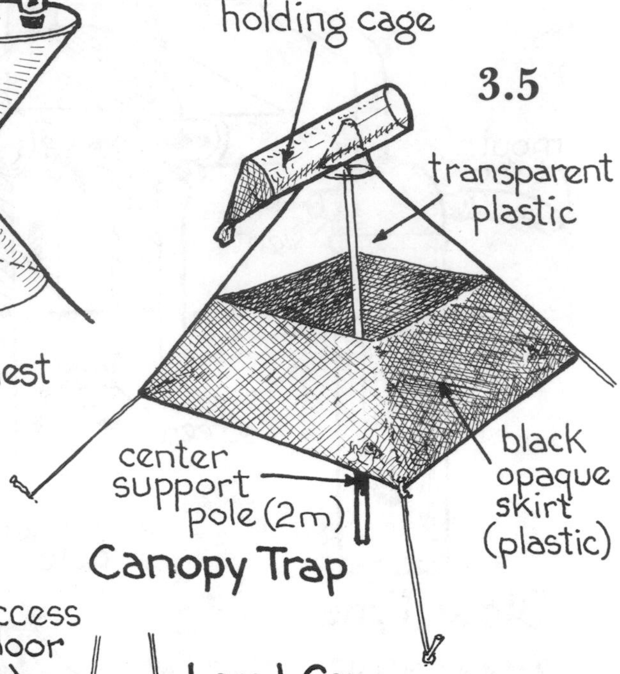

holding cage
3.5
transparent plastic
center support pole (2m)
black opaque skirt (plastic)

Canopy Trap

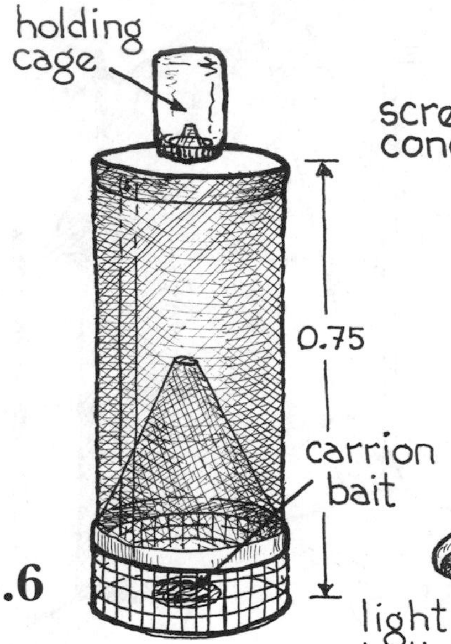

holding cage
screen cone
access door
Lard Can Trap 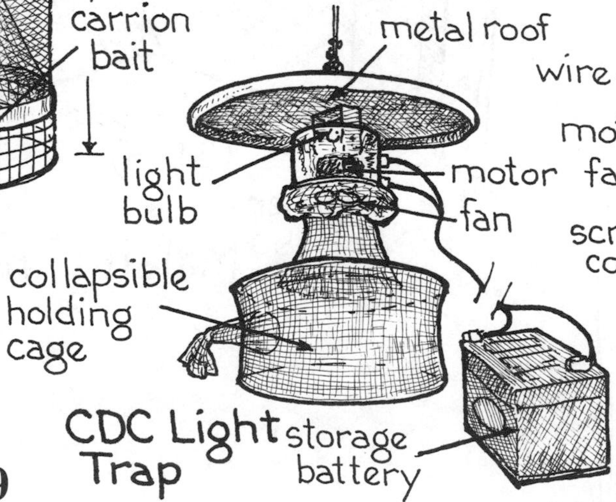
3.7
dry ice
0.75
0.6m

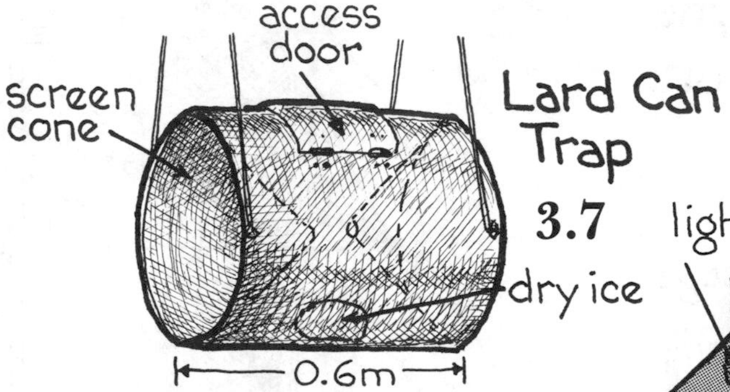

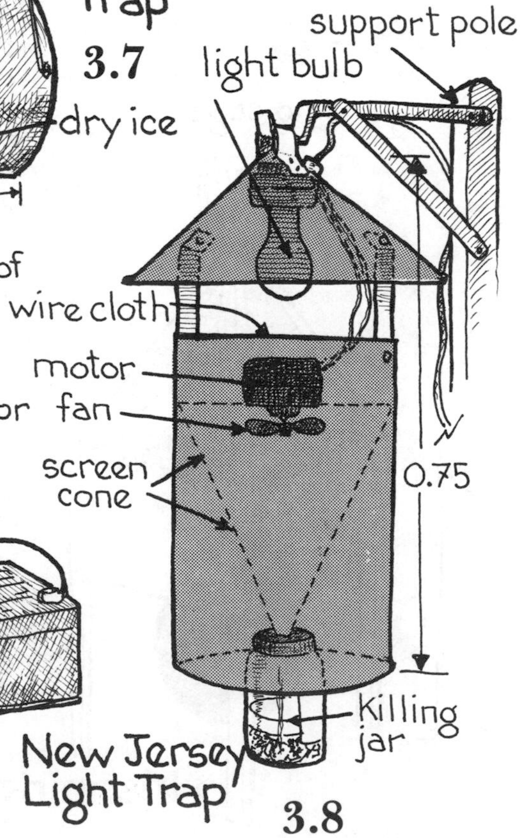

support pole
light bulb
wire cloth
motor fan
screen cone
0.75
killing jar

3.6
carrion bait

Bishopp Fly Trap

metal roof
light bulb
collapsible holding cage
motor fan

CDC Light Trap
storage battery
3.9

New Jersey Light Trap
3.8

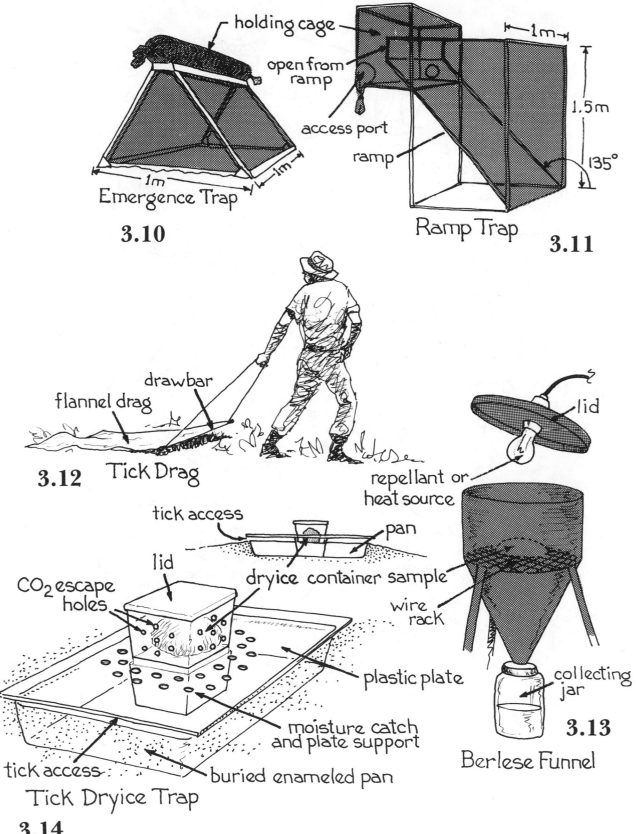

holding cage

open from ramp

access port

ramp

1m

1.5m

135°

1m

Emergence Trap

3.10

Ramp Trap

3.11

drawbar

flannel drag

3.12 Tick Drag

lid

repellant or heat source

tick access

pan

CO_2 escape holes

lid

dryice container sample

wire rack

plastic plate

collecting jar

3.13

moisture catch and plate support

tick access

buried enameled pan

Tick Dryice Trap

Berlese Funnel

3.14

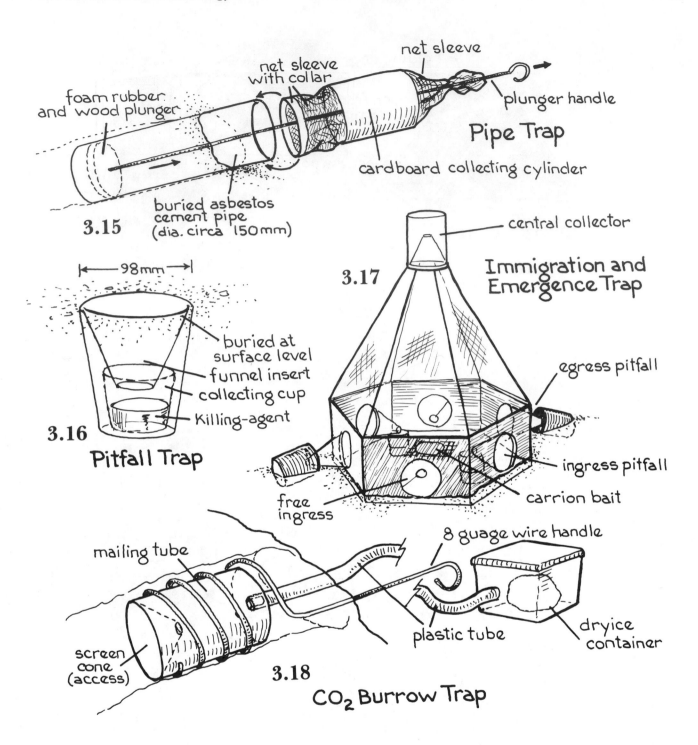

foam rubber
and wood plunger

net sleeve
with collar

net sleeve

plunger handle

Pipe Trap

cardboard collecting cylinder

buried asbestos
cement pipe
(dia. circa 150mm)

3.15

|← 98mm →|

buried at
surface level
funnel insert
collecting cup
Killing-agent

3.16
Pitfall Trap

central collector

3.17

**Immigration and
Emergence Trap**

egress pitfall

ingress pitfall

carrion bait

free
ingress

mailing tube

8 guage wire handle

screen
cone
(access)

3.18
CO_2 Burrow Trap

plastic tube

dryice
container

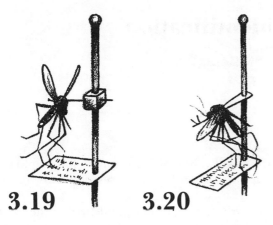

3.19   **3.20**

specimen on minuten
nadeln and cork block

specimen glued
to card point

Flea: mounted on microscope slide
and labeled to show locality data
and identity determination.   **3.21**

specimen glued
to side of pin

3.22

3.23

specimen impaled on
pin and labeled

4   Arthropod taxonomy and identification

Taxonomy

Insect taxonomy (or systematic entomology) is the science of the classification of organisms belonging to the class Insecta. It concerns the placing of insect species in their proper phylogenetic relationships to one another.

Modern systematics arose from the tenth edition of the *Systema Naturae* (1758), published by the great Swedish naturalist Linnaeus. In this work the binomial system of nomenclature used today was established for animal systematics. Based on this system, every animal is given a generic and specific name.

The full name of an insect or any other animal properly includes the author, the person who first named the species; thus the scientific name of the housefly is *Musca domestica* Linnaeus. If the species has since been placed in a genus different from that in which it was originally described, the author's name is placed in parentheses, as *Xenopsylla cheopis* (Rothschild). For the sake of brevity the author's name is omitted in the following keys in the manual.

Note that the generic name is always capitalized, whereas the specific name is not.

The species is the most important taxonomic category. It has been defined as a group of interbreeding natural populations that is reproductively isolated from other such groups. The species thus represents the basic building block with which systematics deals.

A synonym (which is an invalid name for the same category) of the genus or the species may be placed in parentheses after the entire name, and an equals sign should be placed with it, as *Culiseta incidens* (= *Theobaldia*), which indicates that in some older texts the species will be referred to as *Theobaldia incidens*, although we know that it should be in the genus *Culiseta*.

The genus (the plural form of the word is genera) typically is composed of several distinct species showing similarities in structure, although in the absence of close relatives there may be only one species in a group (monotypical). Groups of related genera are classified together as a family, groups of related families as an order, and groups of related orders as a class. Various other intermediate categories are also used. Categories above the species level are intended to show natural relationships and reflect the phylogeny of organisms.

In classification, the system of names used, together with the rules governing their choice, is known as nomenclature (International Code of Zoological Nomenclature).

The generic name is the main foundation of systematic nomenclature. It is unique in that it must be different from every other generic name proposed for an animal. The specific name may occur repeatedly as long as it is found only once in a single genus. Thus names like *californica, maculatus,* and *domestica* occur in many genera. The specific name varies widely in form and length with terminations following only the rules of Latin grammar. Generic names are customarily Latinized names of Greek origin.

The family name always ends in *-idae,* and like all names above the species it is capitalized, as Muscidae. The family name is taken from the stem of the principal genus belonging to the family, as Menoponidae from *Menopon,* and Pulicidae from *Pulex.*

Subfamily names are formed in the same way with the invariable ending *-inae,* thus Culicinae. A common division below the subfamily is the tribe, and this, following the same system, has an *-ini* termination, as in Anophelini.

Names of insect orders are often based on wing structure, in which case the name ends in *-ptera,* as Diptera and Siphonaptera. Ordinal names based on other features usually terminate in *-a, -ea,* or *-ia,* as Mallophaga, Mantodea, and Blattaria.

In some orders we find superfamilies, above the family but below the order. These are always formed from the principal family name, and invariably terminate in *-oidea,* as Muscoidea from Muscidae.

Thus the termination of the scientific name will usually indicate the taxonomic level of the name and the relationship it connotes.

With the aid of the textbook or one of the references supplied, the student will fill in the names of the appropriate taxonomic categories listed below for a representative insect of medical importance, such as *Aedes aegypti* (Linnaeus), or *Musca domestica* Linnaeus.

> Phylum
> Class
> Order
> Superfamily
> Family
> Subfamily
> Tribe
> Genus
> Species

Arthropod identification

The correct identification of the arthropod in hand is of paramount importance. Accurate identification, often at the species level, is needed before control or corrective measures can be taken in most entomological problems. For example, not all mosquito species are vectors of malarial agents, nor are those species that are vectors equally efficient in transmission. Further, each species has its own characteristic larval habitat and adult behavior on which control recommendations must be based. Accurate identification of the problem species therefore gives the medical entomologist the primary key for access to literature dealing with such areas as the distribution, medical importance, ecology, behavior, and control of a species.

Instructions for use of identification keys

Identification keys are based upon the principle of separating species or groups of species by some criterion characteristic of one but not of the other of the categories under consideration. It is customary to split off major groups first, and then successively break up each of these assemblages into smaller groups until finally the basic unit, or species, is reached. Typically, the criterion for separation of the various taxa is an observable morphological character, but host data, distribution data, or other biological information may be used.

Before using a key, first read it through to familiarize yourself with the way it is organized and to make certain you know the meanings of all general terms. Use a dictionary of zoological terms or *A Glossary of Entomology* (Torre-Bueno, 1937) for terms not described or illustrated in the keys. Make a mental note of any descriptions that seem particularly specific, such as in Exercise 14 in the key to *Aedes* mosquitoes (couplet 6, second alternate), "Mesonotum with conspicuous silvery lyre-shaped marking...." This may guide you to an almost immediate identification of the specimen, and the check through the key becomes practically routine.

Where several characters are given in a couplet, read only to the first semicolon; then read the alternative and check that one character before proceeding further. Often the first character suffices to make the distinction; if the character concerned is either not distinctly seen on the specimen or of dubious quality, the other characters given will help you in reaching a decision.

Identification keys should be looked upon as guides, or aids to diagnosis, and absolutely definitive determinations must be made by careful checking with original or revised descriptions of the species or from comparison with authoritatively identified specimens.

With the aid of the keys given below identify to class the arthropods supplied to you for this purpose. Members of the classes Insecta and Arachnida should be identified to order.

Key to classes of arthropods of medical importance

1. Arthropods without wings...2
   • Arthropods with 1 or 2 pairs of thoracic wings, usually well
     developed but rarely reduced or vestigial ....Insecta

2. Arthropods with 3 pairs of thoracic legs and 1 pair of
   antennae . . . . .Insecta
 - Arthropods with only one or neither of the above characters 3

3. Body grub-, caterpillar-, scale-, or saclike or vermiform, with
   tracheal respiration usually evidenced externally by spiracles
   on the abdomen or thorax or both . . . . .Insecta
 - Body without the above combination of characters 4

4. Body wormlike, elongate, flattened or cylindrical, divided into
   a series of conspicuous rings that are not true segments
   (Figures 4.7, 4.8); 2 pairs of fanglike hooks anteriorly on either
   side of mouth; no special respiratory organs (tongue worms
   and allies-adults in respiratory tracts of reptiles, birds, and
   mammals)(considered a subphylum by some authorities) . . . . .Pentastomida
 - Body not with the above combination of characters 5

5. Antennae present 6
 - Antennae absent; body usually with 2 main divisions
   (cephalothorax and abdomen) and 4 pairs of legs (rarely 2 or 3
   pairs, whereas some apparently have 5 pairs due to
   development of pedipalps); compound eyes lacking, simple
   eyes usually present (Figures 19.2, 21.16) (mites, ticks, spiders,
   scorpions, etc.) . . . . .Arachnida

6. Body with 12 or more pairs of walking legs or with 1 pair of
   antennae if with less than 12 pairs of legs (immature forms);
   legs uniramous: respiration tracheal; gills absent . . . . .Myriapoda 7
 - Body with fewer than 12 pairs of walking legs (at least the
   posterior pairs biramous) and usually 2 pairs of antennae (one
   pair reduced or absent in terrestrial forms) (sow bugs, pill
   bugs, crabs, crayfish, shrimp, etc.) . . . . .Crustacea

7. Body with only 1 pair of legs on a body segment; body
   dorsoventrally flattened 8
 - Body with 2 pairs of legs on most segments (millipedes) . . . . .subclass Diplopoda

8. Tarsi with a single claw; spiracles on most segments; first pair
   of legs large, curved, forming poison fangs (centipedes) . . . . .subclass Chilopoda
 - Tarsi with 2 claws; only 1 pair of spiracles, on head; first pair
   of legs short, not modified as poison fangs (symphylans) . . . . .subclass Symphyla

---

Key to orders of insects of medical importance, adults only

1. Wingless, or with vestigial or rudimentary wings 2
 - Winged, though forewings may appear to be part of the body
   due to thickening and chitinization 8

2. Body strongly laterally compressed (Figure 18.1) or
   dorsoventrally depressed (Figures 8.1, 8.4); abdomen more
   or less broadly joined to thorax, not petiolate 3
 - Body cylindrical; abdomen petiolate at base (Figure 21.23)
   (ants and velvet ants) . . . . .Hymenoptera

3. Body dorsoventrally depressed; hind legs not developed for
   jumping. 4
 - Body strongly laterally compressed; hind legs enlarged,
   developed for jumping (fleas) (Figure 18.1) . . . . .Siphonaptera

4. Pronotum small or indistinct, not projecting over head; antennae not more than 5-segmented, short; small (less than 13 mm) 5
   - Pronotum large, projecting over head; antennae many-segmented, long (Figure 5.1); large (13 mm or more) cockroaches) ....Blattaria

5. Tarsi 3- or 5-segmented, usually with 2 large or very small tarsal claws 6
   - Tarsi 1- or 2-segmented or apparently fused to tibiae, usually only a single tarsal claw 7

6. Tarsi slender, 3-segmented, with claws weak, rarely absent; antennae rather long, 4-segmented (bed bugs) (Figure 7.20) ....Hemiptera (Cimicidae)
   - Tarsi usually short, 5-segmented, claws strong, antennae very short and 1-segmented (louse flies) (Figure 2.9) ....Diptera (Hippoboscidae)

7. Piercing-sucking mouthparts; head narrower than thorax (Figure 9.19) (sucking lice) ....Anoplura
   - Chewing mouthparts; head wider than thorax or prolonged into a long proboscis (Figures 8.8, 8.12) (biting lice) ....Mallophaga

8. Forewing entirely membranous, usually quite unlike body in appearance 9
   - Forewing horny, leathery, or only partly membranous, usually closely resembling body in texture and appearance 11

9. Hindwings present, similar to forewings 10
   - Hindwings absent, represented by halteres (Figure 1.3) (flies, mosquitoes, gnats) ....Diptera

10. Mouthparts forming beak, mandibles hidden, styliform; hindwings as large as forewings or nearly so; abdomen broadly joined to thorax (phytophagous; aphids, leafhoppers, cicadas, etc.) ....Homoptera
    - Obvious jawlike mandibles present; hindwings usually considerably smaller than forewings; abdomen petiolate at base (wasps, hornets, bees) ....Hymenoptera

11. Forewings entirely horny or leathery 12
    - Forewings partly membranous, partly thickened (Figure 7.2) (i.e., hemelytra) (phytophagous, predacious, and parasitic true bugs) ....Hemiptera

12. Forewings without true veins, usually meeting in a straight midline, usually horny (Figure 6.14) (i.e., elytra) 13
    - Forewings with distinct veins, folded over each other, leathery (i.e., tegmina) (Figure 5.1) (cockroaches) ....Blattaria

13. Abdomen terminating in a pair of weak or strong pincers (earwigs) ....Dermaptera
    - Abdomen not terminating in a pair of pincers (beetles) ....Coleoptera

Key to orders of Arachnida[a]

1. Abdomen with distinct segmentation 2
   - Abdomen without distinct segmentation (may appear weakly segmented in Ricinulei) 9

2. Abdomen with taillike elongation of more than 5 segments.....................................3
   • Abdomen not elongated or taillike (may have short
     3-segmented tail)...5

3. Elongated portion of abdomen not filamentous, 6-segmented
   with a caudal thornlike terminal stinging apparatus (Figure
   4.1) (scorpions)...order Scorpionida
   • Abdomen elongated into a filamentous or whiplike structure
     of 10 or more segments lacking stinging apparatus
     (Figure 4.2)...4

4. Small, less than 6 mm in length (including tail); caudal
   elongation of abdomen with nodelike segments; agile soil- and
   litter-inhabiting arachnids (micro whip scorpions)..............................order Palpigradi
   • Large (greater than 15 mm in length) to very large (up to 80
     mm exclusive of whiplike tail); caudal elongation cylindrical,
     whiplike (whip scorpions and vinegarones) (Figure 4.2).........................order Uropygi

5. Abdomen with a short, 3-segmented caudal elongation (tail);
   pedipalps nearly twice the length of largest legs, length less
   than 8 mm. World tropics (schizomids)..........................order Schizomida
   • Abdomen without caudal elongation; pedipalps variously
     developed..6

6. Cephalothorax and abdomen appear as a compact, single unit
   (Figure 4.3); 1 pair of dorsomedian eyes (Figure 4.4) podal
   sternites elongated to separate coxae widely; legs usually
   several times as long as length of body (harvestmen or
   daddy-long-legs)...order Opiliones
   • Cephalothorax and abdomen as distinct units (although they
     may be broadly joined); podal sternites variable; legs not
     greatly elongated..7

7. Pedipalps with claws well-developed as pincers (most with
   poison glands); eyes lateral or absent; small size, less than 5
   mm in length; common in leaf litter, manure, moss and under
   loose bark (pseudoscorpions) (Figure 4.5).......................order Pseudoscorpiones
   • Pedipalps without terminal pincers, palpal claws lacking or
     inconspicuous..8

8. First pair of legs without claws, very long (much longer than all
   other legs) and slender, whiplike; abdomen attached to
   cephalothorax by a narrow stalk; cephalothorax wider than
   abdomen; moderate size, 10–15 mm in length (tailless whip
   scorpions) (one family, Tarantulidae)...........................order Amblypygi
   • First pair of legs with or without claws, subequal in length and
     diameter to at least some legs; cephalothorax broadly joined to
     abdomen; chelicerae large and pincerlike (wind scorpions or
     sun scorpions) (Figure 4.6)..................................order Solifugae

9. Abdomen constricted to form a narrow waist (Figure 21.16)
   (spiders) (Exercise 23)......................................order Araneida
   • Abdomen broadly joined to cephalothorax
     (Figures 19.1, 20.1)...10

10. Abdomen appears as a distinct unit separate from
    cephalothorax, head with hinged hood that covers chelicerae
    in repose; size 10–15 mm in length (ricinuleids)..............order Ricinulei

- Separation of cephalothorax and abdomen indistinct; size generally small (3 mm or less except for suborder Ixodida, or Metastigmata, with size up to 30 mm); body habitus variable (mites and ticks) (Exercises 19 and 20) ....order Acarina
(= subclass Acari of authors)

*a*Modified from Scott and Stojanovich (1966). Krantz (1978) raises these orders to subclasses.

References

Borror, D. J., D. M. DeLong, and C. A. Triplehorn. 1976. *An Introduction to the Study of Insects,* 4th ed. Holt, Rinehart and Winston, New York.

Brues, C. T., A. L. Melander, and F. M. Carpenter. 1954. Classification of insects. *Bull. Mus. Comp. Zool. Harv. Univ.* 73:1–917.

Comstock, J. H. 1940. *An Introduction to Entomology,* 9th ed. Comstock, Ithaca, N.Y.

Daly, H. V., J. T. Doyen, and P. R. Ehrlich. 1978. *Introduction to Insect Biology and Diversity.* McGraw-Hill, New York.

Essig, E. O. 1958. *Insects and Mites of Western North America,* 2nd ed. Macmillan, New York.

– 1942. *College Entomology.* Macmillan, New York.

Faust, E. C., P. F. Russell, and R. C. Jung. 1970. *Craig and Fausts' Clinical Parasitology,* 8th ed. Lea & Febiger, Philadelphia.

Foote, R. H. 1977. *Thesaurus of Entomology.* Entomological Society of America, College Park, Md. 188 pp.

Imms, A. D., revised by Richards, O. W. and R. G. Davies. 1957. *A General Textbook of Entomology,* 9th ed. Methuen, London.

International Commission on Zoological Nomenclature. 1961. *International Code of Zoological Nomenclature.* International Trust for Zoological Nomenclature, London.

Kaston, B. J. 1978. *How to Know the Spiders,* 3rd ed. The Picture Key Nature Series. Wm. C. Brown, Dubuque, Iowa. 272 pp.

Krantz, G. W. 1978. *A Manual of Acarology,* 2nd ed. Oregon State University Book Stores, Corvallis.

Levi, H. W., and L. R. Levi. 1968. *A Guide to Spiders and Their Kin.* A Golden Nature Guide. Golden Press, New York. 160 pp.

Mayr, E. 1969. *Principles of Systematic Zoology.* McGraw-Hill, New York.

Metcalf, C. L., W. P. Flint, and R. L. Metcalf. 1962. *Destructive and Useful Insects,* 4th ed. McGraw-Hill, New York.

Ross, H. H. 1965. *A Textbook of Entomology,* 3rd ed. Wiley, New York.

Scott, H. G., and C. J. Stojanovich, 1966. *Pictorial Keys: Archropods, Reptiles, Birds and Mammals of Public Health Significance.* U.S. Dept. of Health, Education, and Welfare, USPHS, Washington, D.C. 192 pp.

Torre-Bueno, J. R. de la. 1937, 1960. *A Glossary of Entomology.* Brooklyn Entomological Society, New York. 336 pp., 9 plates.

Tulloch, G. S. 1960. *Torre-Bueno's Glossary of Entomology,* suppl. A. Brooklyn Entomological Society, New York. 36 pp.

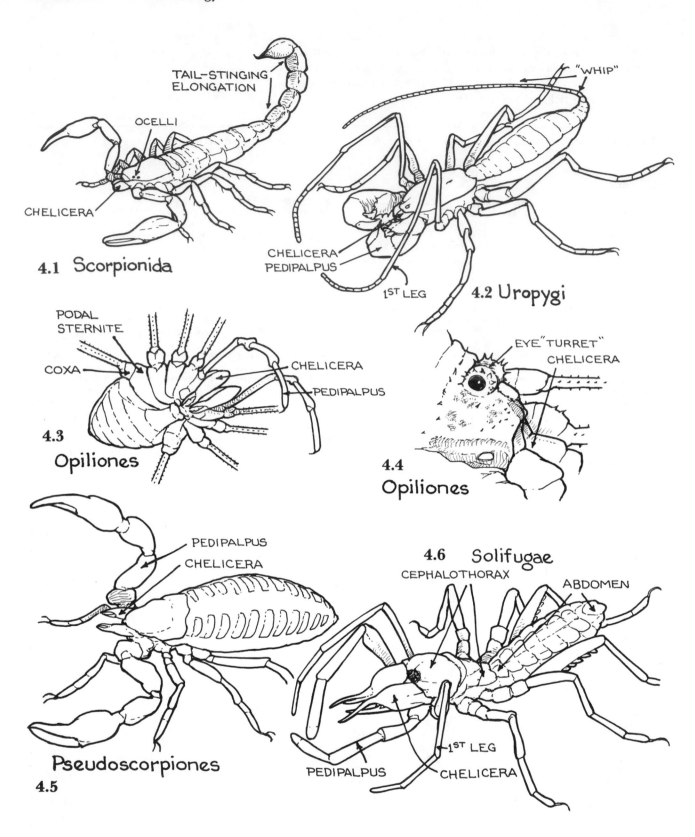

4.1   Scorpionida

4.2 Uropygi

4.3 Opiliones

4.4 Opiliones

4.6   Solifugae

Pseudoscorpiones
4.5

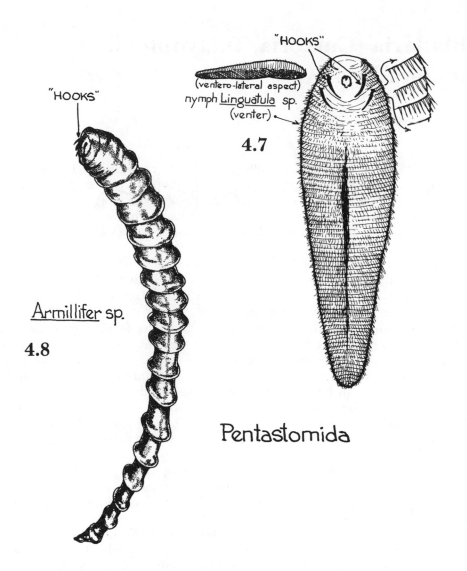

"HOOKS"

(ventero-lateral aspect)
nymph _Linguatula_ sp.
(venter)

4.7

"HOOKS"

Armillifer sp.

4.8

Pentastomida

5 Order Blattaria (Cursoria, Dictyoptera, or Orthoptera in part)

These are the cockroaches, a very old and morphologically generalized group of insects having gradual metamorphosis. They are clearly adapted for cursorial behavior, and although wings are well-developed in many species, most fly rather infrequently. Because of their omnivorous diet, gregarious nocturnal behavior, and house-frequenting occurrence, some species are in close contact with humans and are associated with a wide spectrum of disease agents. These disease relationships have been reviewed in detail by Roth and Willis (1957).

Cockroaches will eat practically every human food as well as human wastes – whether garbage, vomit, or excrement. Disease agents are usually transmitted by cockroaches mechanically by their regurgitations or through fecal or surface contamination of human food. In the transmission of certain helminth parasites of animals, cockroaches act as a necessary biological intermediate host.

More than 3,500 species of cockroaches are known. They are placed in five families generally in accordance with the manner in which the eggs are brooded and the young are produced. Many species package their eggs in a purselike structure called an **oötheca.** Others are ovoviviparous. Most cockroaches are tropical forest inhabitants. Less than 1% of the known species are encountered commonly in human dwellings. The few species that do thrive in buildings are cosmopolitan in distribution. Rehn (1945) describes the probable mechanisms for the worldwide dispersal of the dozen more common house-frequenting cockroaches. Other species, such as wood or field cockroaches, may be carried into houses on firewood, potted plants, garden produce, or other objects. A few species will fly to lights at night. Generally such introductions do not survive for long in houses.

In the following key most house-frequenting species are included along with representatives of wood or field cockroaches commonly introduced into homes. Others of the 55 species that occur in the United States (exclusive of Hawaii) can be identified using Helfer's keys (1963). The 18 species found in Hawaii can be identified by referring to Zimmerman (1948).

For keying, cockroaches are best kept in liquid preservative, such as formalin or 70% ethyl alcohol, to allow manipulation of parts for examination. If only pinned specimens are available, they can be softened readily by submerging them for a few hours in a warm 0.5% aqueous solution of tri-sodium phosphate or another similar wetting agent.

The purpose of this exercise is twofold: to enable students to key and recognize common house-frequenting cockroaches and to expose students to the use of a key for discriminating between similar-appearing species. The key that follows is for identifying adult cockroaches only, and in many instances the sex also must be determined. Immature cockroaches, or nymphs, closely resemble the adults except that they are generally smaller and lack wings and genital openings or copulatory appendages at the tip of the abdomen. In some species the adult females also are wingless, but nymphs can be recognized by their shorter, relatively broad cerci and lack of external genitalia. Male cockroaches possess styli in addition to the paired cerci. The styli arise from the subgenital plate and are conspicuous in most species, but they may be reduced in some. Styli are absent in females and nymphs.

Note that the leathery, protective front wing of cockroaches is termed the **tegmina,** and the hind wing is the membranous **wing.**

Key to some house-frequenting and other more common species of Blattaria, adults[a]

1. Wing with an intercalated triangle (or appendicular area) large (Figures 5.2, 5.3) 2
 • Wing without such a triangle, or with area inconspicuous (Figure 5.12) 5

2. Profemur with 3 midventral spines, 2 apical spines; overall color light yellow brown with many darker spots; length 8–10 mm (range: Africa, Europe, and Massachusetts) Spotted Mediterranean cockroach . . . .*Ectobius pallidus*

   ● Profemur without midventral spines, with only a fringe of hairs and 1 or 2 apical spines 3

3. Tegmina with many veins; appendicular wing area about one-sixth of wing surface (Figure 5.2); length 7–9 mm (range: Texas through southeastern U.S.) Small Texas cockroach . . . .*Chorisoneura texansis*

   ● Tegmina with few conspicuous veins; appendicular wing area large (one-third or more of wing surface) (Figure 5.3) 4

4. Habitus small, oval, convex; length 5–6 mm (range: Florida, Key West, and Caribbean) Florida beetle cockroach . . . .*Plectoptera poeyi*

   ● Habitus large, elongate, depressed; length 35–40 mm (range Pacific and Orient) Cypress cockroach . . . .*Diploptera dytiscoides*

5. Profemur with spines along ventral margin and 1 or more apical spines 6

   ● Profemur without marginal spines (may have fringe of 30 or more hairs) 14

6. Tegmina absent; body hairy; protibia with 7 apical spines; mid and hind femora without spines but very hairy; length 8–13 mm (range: deserts of southwestern U.S.; attracted to lights) Hairy desert cockroach . . . .*Eremoblatta subdiaphana*

   ● Tegmina present (may be reduced and/or fused); body not usually hairy 7

7. Profemur with spines subequal or of gradual difference in length (Figure 5.4) (do not confuse spines with hairs) .8

   ● Profemur with spines of distinctly different size classes (Figures 5.5, 5.6) 9

8. Pronotum and abdomen with numerous distinct yellow markings; tegmina much reduced; length 20–25 mm (range: tropical Asia, Pacific, and Mexico) Harlequin cockroach . . . .*Neostylopyga rhombifolia*

   ● Pronotum and abdomen not so marked 20

9. Pronotum, tegmina, and abdomen strikingly marked (Figure 5.8); 4th tarsal segment only with pulvillus; length 7–10 mm (range: Texas to Florida) Little gem cockroach . . . .*Aglaopteryx gemma*

   ● Tegmina and abdomen unmarked; basal 4 tarsal segments with pulvilli (Figure 5.6) 10

10. Pronotum with two longitudinal dark bars (Figure 5.1); styli reduced and globular (Figure 5.9); length 10–15 mm .11

    ● Pronotum not so marked; if styli reduced, not globular 12

11. Male virga grooved distally (Figure 5.9); aedeagus (phallomere) not swollen distally (may be necessary to dissect out aedeagus for examination) (range: cosmopolitan) German cockroach . . . .*Blattella germanica*

- Male virga smooth distally; aedeagus swollen distally (range: Orient and Pacific tropics; a field cockroach)
  False German cockroach . . . .*Blattella lituricollis*

12. Male with tergites VI and VII specialized in the middle; median segment not specialized; 1 style medially placed, nearly double the other in size (Figure 5.10) (range mid- and southeastern U.S.; a woodland species)
    Dark wood cockroach . . . .*Ischnoptera deropeltiformis*
- Male without specialized tergites VI and VII, but often with median segment specialized; styli subequal in size 13

13. Wing with mediastine vein reaching near apex (Figure 5.11); male median segment unspecialized; sexes similar; length 15–20 mm (range: tropical)
    Pallid cockroach . . . .*Leurolestes pallidus*
- Wing with mediastine vein reaching about halfway to apex (Figure 5.12); female usually with reduced tegmina and wings; male often with specialized median segment (Figure 5.13), length 9–22 mm (genus occurs throughout the U.S.)
  Wood cockroaches . . . .*Parcoblatta* spp.

14. Habitus a general pale green color; sexes similar; top of eyes close together; length 12–18 mm (range: Texas, Mexico, Central America, Greater Antilles)
    Cuban cockroach . . . .*Panchlora nivea*
- Habitus not colored as above 15

15. Large body size (more than 35 mm long); ventral (anterior) margins of mid and hind femora with 1 short apical spine; profemur with only a fringe of hairs (on the ventral margin); no spine; tip of male abdomen shieldlike (Figure 5.14); length 38–53 mm (range: West Africa, Central and S. America, tropical Pacific, eastern U.S.)
    Maderan cockroach . . . .*Leucophaea maderae*
- Smaller, length less than 30 mm 16

16. Habitus oval, small; length 8–13 mm 17
- Habitus elongate, medium size; length 16–30 mm 18

17. Each black tegmina with a large, sharply defined round orange-yellow spot; pronotum black, fringed with long erect hairs, with yellow spot on posterolateral margin; length 10 mm (range: tropicopolitan)
    Pacific beetle cockroach . . . .*Euthyrrhapha pacifica*
- Tegmina and pronotum not so marked; body densely clothed with yellowish hairs; tegmina present in male, not in female; a desert species; length 8–13 mm (range: deserts of southwestern U.S.)
  Hairy desert cockroach . . . .*Eremoblatta subdiaphana*

18. Pronotum with front edge thickened and projected to form a hood over the head; tegmina absent in both sexes; profemur with 3 strong distal spines; length 23–30 mm (range: most of U.S., occurring in wet, rotten wood)
    Brown hooded cockroach . . . .*Cryptocercus punctulatus*
- Pronotum without thickened or projected front edge 19

19. Pronotum dark brown with pale lateral and anterior margins;
    tegmina light brown, basal half of discoidal vein dark brown,
    basal fourth with rows of small round pits (Figure 5.15);
    profemur with border of fine spines and 1 stout apical spine;
    length 16–24 mm (range: world tropics)
    Burrowing or Surinam cockroach .... *Pycnoscelis surinamensis*
- Pronotum pale brown, with central blotch having intricate
    markings, a pale lateral margin, and a darker sublateral
    border; basal third of tegminal discoidal vein brown (Figure
    5.16); length 20–25 mm (range: world tropics)
    Cinereous cockroach .... *Nauphoeta cinerea*

20. Female with sternite VII divided (Figure 5.17) to form a
    valvular apparatus; size medium to large, longer than 16 mm
    (Blattidae) .... .21
- Sternite VII of female large, undivided and rounded (Figure
    5.7), size very large (40–80 mm) or small (10–20 mm) .... .26

21. Tegmina well developed, nearly covering abdomen or beyond;
    size medium to large, over 23 mm in length .... *Periplaneta* spp. 23
- Tegmina reduced (covering two-thirds or less of abdomen);
    length 15–40 mm .... .22

22. Tegmina short, wider than long; hind tarsus with basal
    segment shorter than combined length of other segments;
    pulvilli of segments II and III large (Figure 5.20); length
    30–40 mm (range: southeastern U.S.)
    Stinking cockroach .... *Eurycotis floridana*
- Tegmina of female small, widely separated, leaving 2 or more
    basal abdominal segments exposed; male tegmina covering
    two-thirds of abdomen; hind tarsus of both sexes with basal
    segment longer than other segments combined (Figure 5.21);
    length 15–25 mm (range: cosmopolitan)
    Oriental cockroach .... *Blatta orientalis*

23. Color above shining blackish brown; length 24–33 mm (range:
    southern U.S.)
    Smoky brown cockroach .... *Periplaneta fuliginosa*
- Color not as above; with distinct to vague pronotal markings .... .24

24. Tegmina with conspicuous lateral pale basal stripe; pronotum
    with sharply contrasting, pale or yellow margin (Figure 5.22);
    length 23–29 mm (range: world tropics and in greenhouses)
    Australian cockroach .... *Periplaneta australasiae*
- Tegmina entirely reddish brown; pronotum with less defined
    markings .... .25

25. Distal segment of cercus elongated, length more than 2×
    width; male with caudal tergite deeply notched (Figure 5.18),
    distal portion of plate thin, projecting as a hood over
    corresponding terminal sternite; median segment
    unspecialized; length 23–35 mm (range: cosmopolitan)
    American cockroach .... *Periplaneta americana*
- Distal segment of cercus triangular, length less than 2× width;
    male with caudal tergite only slightly notched (Figure 5.19),
    distal portion opaque; median segment specialized (see Figure
    5.13), consisting of a shallow channel having a tuft of hairs;
    length 25–33 mm (range: world tropics, southern California)
    Southern brown cockroach .... *Periplaneta brunnea*

26. Pronotum distinctly wider than long; profemur with a few mid
    spines and 1 apical spine; overall outline somewhat oval; very
    large, length 40–80 mm (Blaberidae) 27
  - Pronotum width and length subequal; profemur with
    numerous marginal spines; overall outline elliptical; legs
    slender; length 10–15 mm (Phyllodromiidae) 28

27. Overall color brownish; pronotum dull yellow with a central
    black-brown marking having lighter central area that suggests
    a smiling human face (Figure 5.23); length 40–60 mm (range:
    tropical America)
    Death's head cockroach ....*Blaberus craniifer*
  - Overall color yellowish; tegmina with central brown blotch;
    pronotum dull yellow with clearly defined, shield-shaped
    black-brown central area; length 50–80 mm (range: tropical
    America)
    Giant cockroach ....*Blaberus giganteus*

28. Pronotum with two longitudinal dark bars (Figure 5.1) ....(See couplet 11)
  - Pronotum with a darker central area but without longitudinal
    bars; tegmina darker brown basally, fading toward apex, a
    lighter, sometimes ill-defined, band on the basal half (Figure
    5.24) (range: world tropics and many temperate areas; flies
    readily, attracted to lights)
    Brown-banded cockroach ....*Supella longipalpa*

---

[a]Modified from Helfer (1963).

References

The general textbooks previously listed.

Gould, G. E., and H. O. Deay. 1940. *The Biology of Six Species of Cockroaches Which Inhabit Buildings*. Purdue Univ. Agr. Exp. Sta. Bull. 451. Purdue University, Lafayette, Ind.

Helfer, J. R. 1963. *How to Know the Cockroaches and Their Allies*. Pictured-Key Nature Series. Wm. C. Brown, Dubuque, Iowa.

McKittrick, F. A., 1964. *Evolutionary Studies of Cockroaches*. Memoir 389, Cornell Univ. Agr. Exp. Sta. Cornell University, Ithaca, N.Y.

Rehn, J. A. G. 1945. Man's uninvited fellow traveler – the cockroach. *Sci. Monthly* 61:265–76.

Roth, L. M., and E. R. Willis. 1957. *The Medical and Veterinary Importance of Cockroaches*. Smithsonian Misc. Collection 134(10). 147 pp., 7 plates.

Zimmerman, E. C. 1948. *Insects of Hawaii*, vol. 2, *Apterygota to Thysanoptera*. University of Hawaii Press, Honolulu. 475 pp.

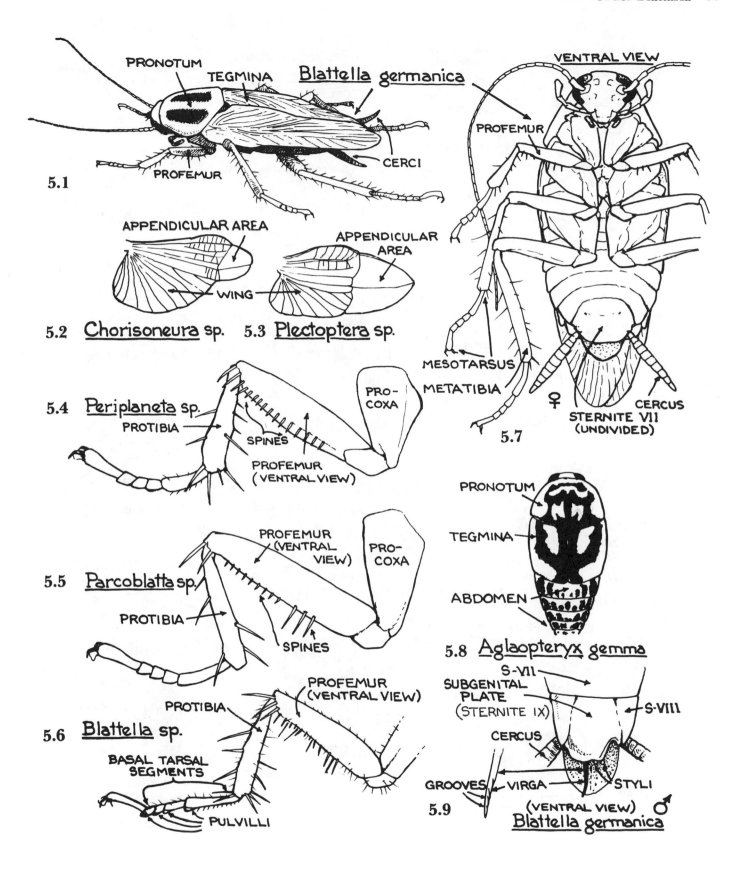

5.1 PRONOTUM TEGMINA <u>Blattella germanica</u>
PROFEMUR
CERCI

VENTRAL VIEW
PROFEMUR
MESOTARSUS
METATIBIA
♀
STERNITE VII
(UNDIVIDED)
CERCUS
5.7

APPENDICULAR AREA
WING
APPENDICULAR AREA

5.2 <u>Chorisoneura</u> sp.   5.3 <u>Plectoptera</u> sp.

5.4 <u>Periplaneta</u> sp.
PROTIBIA
SPINES
PROFEMUR
(VENTRAL VIEW)
PRO-COXA

5.5 <u>Parcoblatta</u> sp.
PROFEMUR
(VENTRAL VIEW)
PRO-COXA
PROTIBIA
SPINES

PRONOTUM
TEGMINA
ABDOMEN
5.8 <u>Aglaopteryx gemma</u>

5.6 <u>Blattella</u> sp.
PROTIBIA
PROFEMUR
(VENTRAL VIEW)
BASAL TARSAL SEGMENTS
PULVILLI

S-VII
SUBGENITAL PLATE
(STERNITE IX)
S-VIII
CERCUS
GROOVES
VIRGA
STYLI
5.9
(VENTRAL VIEW) ♂
<u>Blattella germanica</u>

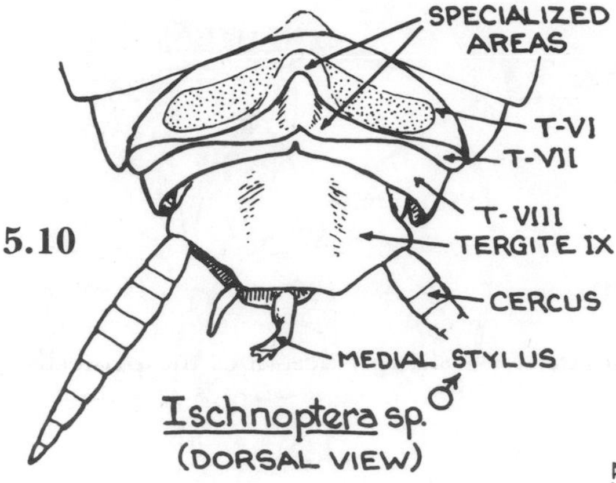

5.10 Ischnoptera sp. (DORSAL VIEW)

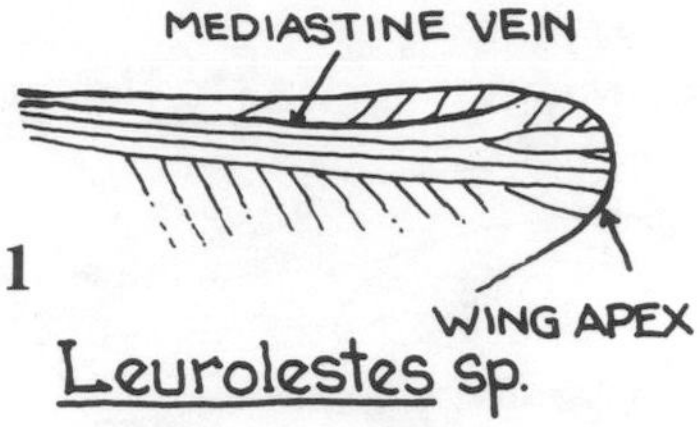

SPECIALIZED AREAS
T-VI
T-VII
T-VIII
TERGITE IX
CERCUS
MEDIAL STYLUS

5.11 Leurolestes sp.

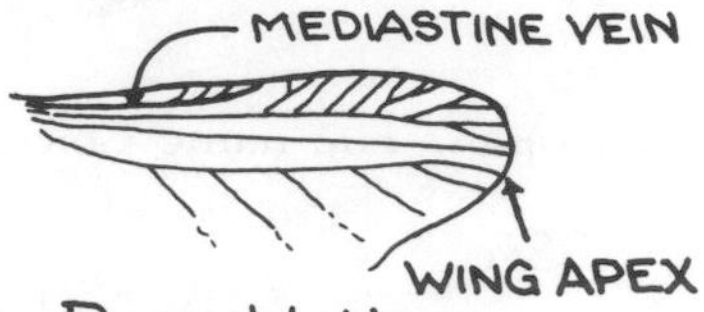

MEDIASTINE VEIN
WING APEX

5.12 Parcoblatta sp.

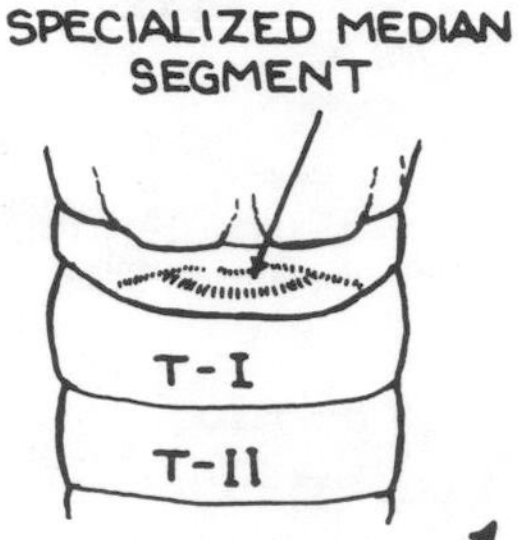

MEDIASTINE VEIN
WING APEX

5.13 Parcoblatta sp. ♂

SPECIALIZED MEDIAN SEGMENT
T-I
T-II

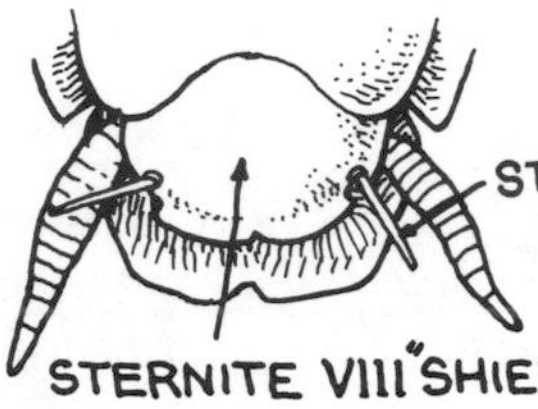

5.14 Leucophaea maderae ♂

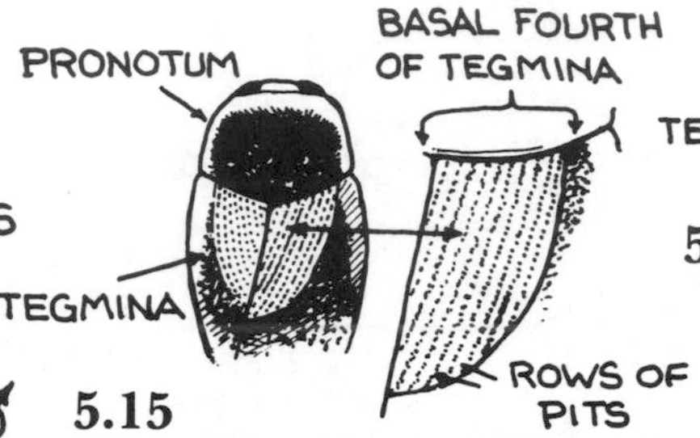

STYLUS
STERNITE VIII "SHIELD"

5.15 Pycnoscelis surinamensis

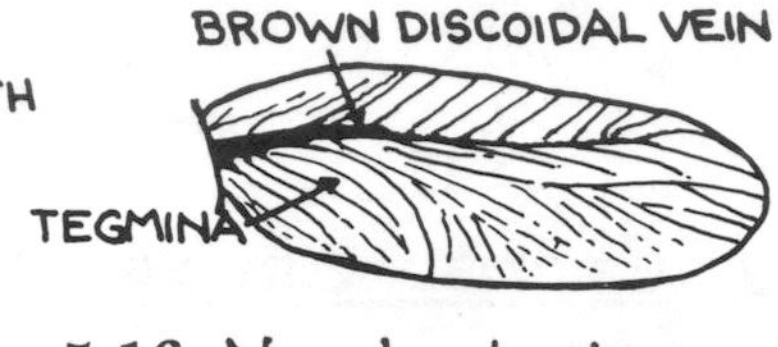

PRONOTUM
BASAL FOURTH OF TEGMINA
TEGMINA
ROWS OF PITS

5.16 Nauphoeta cinerea

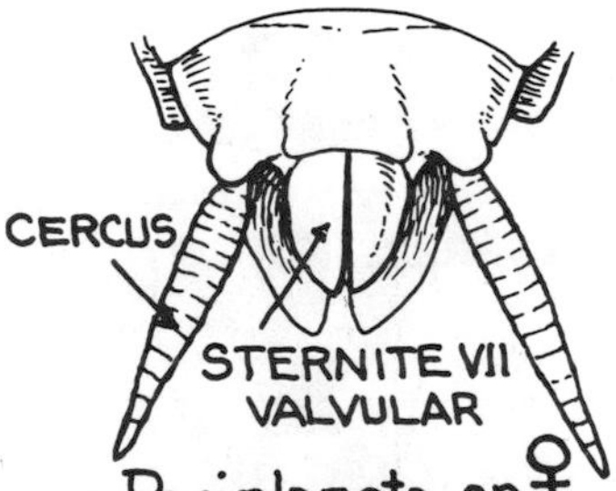

BROWN DISCOIDAL VEIN
TEGMINA

5.17 Periplaneta sp. ♀

CERCUS
STERNITE VII VALVULAR

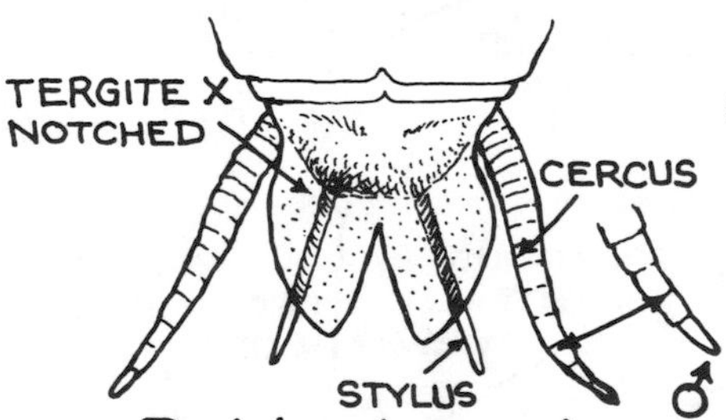

5.18 Periplaneta americana

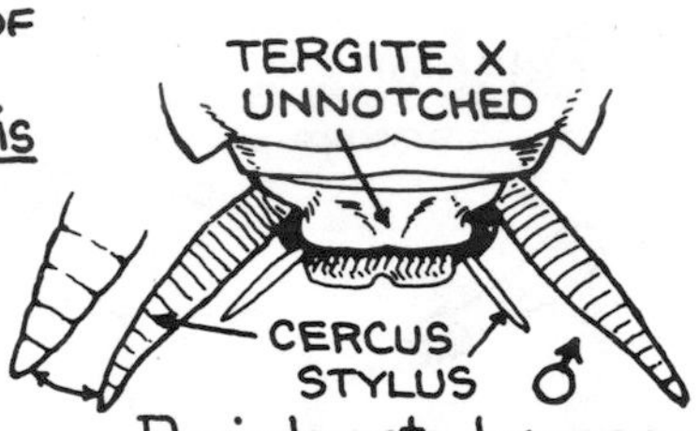

TERGITE X NOTCHED
CERCUS
STYLUS ♂

5.19 Periplaneta brunnea

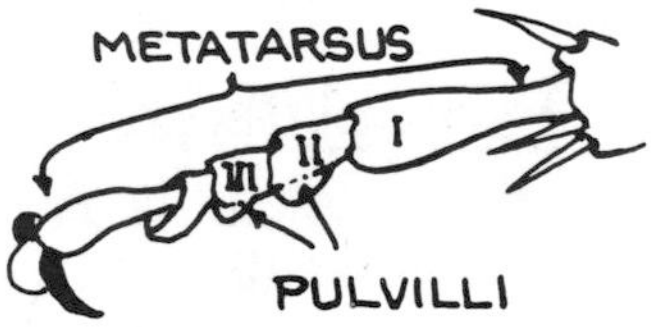

TERGITE X UNNOTCHED
CERCUS
STYLUS ♂

5.20 Eurycotis sp.

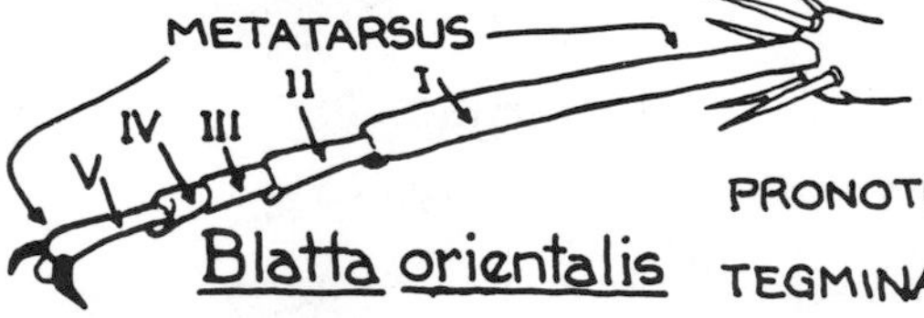

METATARSUS
III II I
PULVILLI

5.21 Blatta orientalis

METATARSUS
V IV III II I

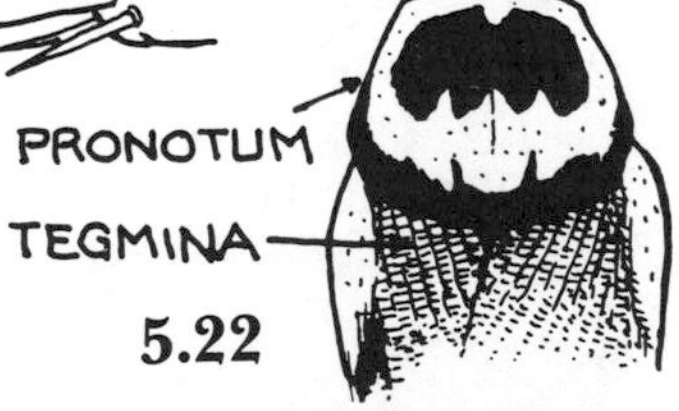

5.22 Periplaneta australasiae

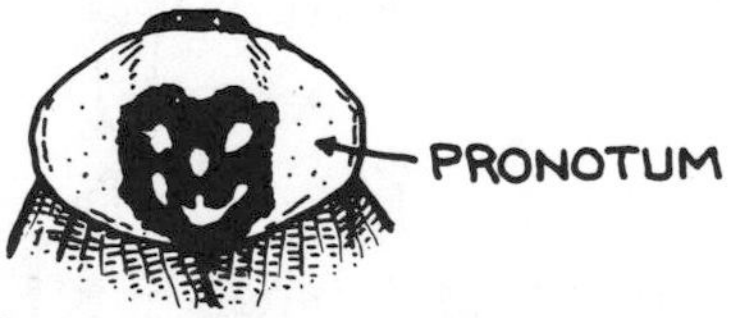

PRONOTUM
TEGMINA

5.23 Blaberus craniifer

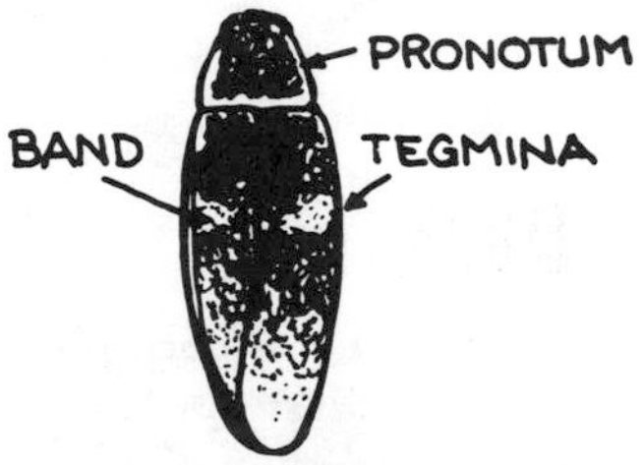

PRONOTUM

5.24 Supella longipalpa

BAND
PRONOTUM
TEGMINA

6   Order Coleoptera

The order Coleoptera, the beetles, is the largest in number of species of all such animal groups. They range in size from minute feather-winged beetles (Ptiliidae) to the giant scarab (Scarabaeidae) and long-horned (Cerambycidae) beetles of the tropics. The name Coleoptera (*Coleos* = sheath) describes the protecting, leathery, sheathlike forewing (mesothorax), termed the **elytra.** Beetles possess strong chewing-type mouthparts, and most members have five-segmented tarsi. The tarsal "formula" is indicated as 5-5-5, each number representing the number of tarsal segments for the first, second, and third pairs of legs, respectively.

Beetles are a major arthropod element in terrestrial and nonmarine aquatic communities. They have complex, or complete, metamorphosis, permitting the larva (grub) to exploit an ecological niche completely different from that of the adult. There is a wide range of larval forms, reflecting the diversity of niches occupied. Adults also display a diversity of forms, ranging from the more generalized cursorial type (Figure 6.1) to those that are specialized (Figures 6.15, 6.16).

Coleoptera are of relatively little importance to medical/veterinary entomology, but because of the ecological diversity of immatures and adults they are of interest for a variety of reasons. Many members of the order serve as intermediate hosts of helminths parasitic in vertebrates. Accidental invasion of natural body openings by beetles is not unusual. Scavenger and coprophagous or carrion-frequenting species may transmit pathogens through contamination. Certain beetles cause vesication (blistering) upon contact, and many can cause intense irritation when contacting the eyes. Obligatory symbiosis is poorly represented in this large group; a few species in the families Leptinidae and Staphylinidae are either inquilines or ecto-symbionts of some marsupials and rodents.

There are more than 100 families of beetles. This key includes only those beetle families and a few genera that are most likely to be encountered by the medical/veterinary entomologist. In addition several groups are included that are stored-products pests often encountered in human dwellings or farm and storage buildings. More complete keys to Coleoptera can be found in most general entomology texts (see References, Exercise 4).

Key to selected families and genera of Coleoptera

1. Pleural sutures of prothorax present; outer lobe or galea of maxilla palpiform (Figure 6.4); first visible abdominal sternite divided by hind coxal cavities (Figures 6.1, 6.3) (predacious ground beetles) ....Carabidae
   - Pleural sutures of prothorax absent; outer lobe or galea of maxilla not palpiform; first visible abdominal sternite not usually interrupted by the hind coxal cavities 2

2. Antennae without an apical comblike club 3
   - Antennae with apical 4–7 segments enlarged on one side to form a comblike or lamellate club (dung beetles, cockchafers, rosechafers, June beetles, scarabs) (Figures. 6.5, 6.6) ....Scarabaeidae

3. Hind tarsi 4-segmented, front and middle tarsi 5-segmented 4
   - Hind tarsi with at least as many segments as the others 6

4. Front coxal cavities open behind (Figure 6.10) 5
   - Front coxal cavities closed behind (darkling ground beetles) (Figures 6.11, 6.14) ....Tenebrionidae

5. Head gradually narrowed behind the eyes (Figure 6.7); slender, soft-bodied beetles, some causing blisters on contact with man . . . . .Oedemeridae
  - Head strongly and suddenly constricted proximally (Figures 6.8, 6.9) (blister and flower beetles) . . . . .Meloidae

6. Prosternum with a median process projecting backward and fitting into mesosternum (Figure 6.2) 7
  - Prosternum without such a process 8

7. The hind angles of prothorax backwardly produced; first and second abdominal segments not fused; free and movable; median process of prosternum not fused to mesosternum (click beetles) . . . . .Elateridae
  - Hind angles of prothorax not strongly produced; first and second abdominal segments fused or immovable; median process of prosternum fused with mesosternum (metallic wood borers – members of genus *Melanophila* – attracted to smoke, may bite skin of man) . . . . .Buprestidae

8. Front coxae conical and projecting; hind coxae dilated into plates partly covering bases of femora (Figure 6.12) (skin, larder, and carpet beetles) . . . . .Dermestidae 17
  - Coxae not as above 9

9. Elytra very short, exposing more than half the abdomen dorsally, or, if more than half the abdomen covered, the eyes absent 10
  - Elytra long, almost or entirely covering the abdomen 11

10. Eyes present; greater part of abdomen exposed dorsally (Figure 6.13); wings present and folded under elytra when not in use (rove beetles and parasitic beetles on rodents, marsupials) . . . . .Staphylinidae
  - Eyes absent; 5 abdominal tergites exposed; wings absent (Figure 6.15) (the beaver parasite) . . . . .Leptinidae (in part) (= Platypsyllidae), *Platypsyllus castoris*

11. Eyes well developed; small to large species 12
  - Eyes absent or greatly reduced; very small species associated with rodents and their nests . . . . .Leptinidae

12. Abdomen with 6 visible ventral segments 13
  - Abdomen with 5 visible ventral segments 14

13. Large (over 12 mm long) flattened or heavy-set beetles, the latter with elytra short, exposing last 2 or 3 abdominal segments (carrion beetles) . . . . .Silphidae 19
  - Smaller and more or less cylindrical in body form; elytra covering abdomen dorsally (copra or ham beetles) . . . . .Corynetidae 20

14. Front coxae longer than broad 15
  - Front coxae broader than long (dried fruit beetles) . . . . .Nitidulidae

15. Antennae terminating in a number of toothlike or serrate segments 16
  - Antennae strongly clavate or capitate but not abruptly so (stored-food product beetles) . . . . .Silvanidae 21

16. Antennae inserted on the front of the head, close together at the base (Figure 6.16) (spider beetles, stored animal and vegetable product beetles) ....Ptinidae
* Antennae inserted on the sides of the head in front of the eyes and usually widely separated at the base (death watch, drug store, and cigarette beetles) ....Anobiidae

Dermestidae

17. With ocellus; smaller (approximately 6 mm and under) 18
* Without ocellus; larger (6 mm and over) ....*Dermestes*

18. Body clothed with short hairs; last antennal segment much the longest; metacoxal plate extending laterally halfway across the parapleura ....*Attagenus*
* Body clothed with short scales; metacoxal plate extending only to the inner boundary of the parapleura ....*Anthrenus*

Silphidae

19. Antennae 11-segmented, slender or gradually clavate (Figure 6.17); body usually somewhat oval, somewhat depressed ....*Silpha*
* Antennae 10-segmented, capitate (Figure 6.18), the last 4 segments forming an abrupt club; body somewhat oblong, usually robust ....*Necrophorus*

Corynetidae

20. Last palpal segment triangular, especially of the labial palpus ....*Corynetes*
* Last palpal segment oval, blunt ....*Necrobia*

Silvanidae

21. Body more or less oval or oblong 22
* Body with sides appearing distinctly parallel (grain beetles) ....*Cathartus*

22. Sides of pronotum with 6 short teeth (sawtoothed grain beetle) ....*Oryzaephilus*
* Sides of pronotum simple or crenulate ....*Silvanus*

References

The general textbooks previously listed.

Arnett, R. H., Jr. 1960-2. *The Beetles of the United States.* Catholic University of America Press, Washington, D. C.

Bradley, J. C. 1930. *Genera of Beetles of America North of Mexico.* Daw, Illston, Ithaca, N.Y.

Hall, M. C. 1929. *Arthropods as Intermediate Hosts of Helminths.* Smithsonian Misc. Collection, 81(15).

Neveu-Lemaire, M. 1933. Les arthropodes hôtes intermediares des helminthes parasites de l'homme. *Ann. de Parasitol. Humaine et Comparée.* 11:222-37, 303-19, 370-97.

– 1938. *Traite d'entomologie medicale et veterinaire.* Vigot Frères, Paris.

Riley, Wm. A., and Johannsen, O. A. 1938. *Medical Entomology,* 2nd ed. McGraw-Hill, New York.

Seevers, C. H. 1955. A revision of the tribe Amblyopinini; staphylinid beetles parasitic on mammals. *Fieldiana. Zoology* 37:211-64, Chicago Natural History Museum.

Theodorides, J. 1950. The parasitological, medical and veterinary importance of Coleoptera. *Acta Trop.* 7:48-60.

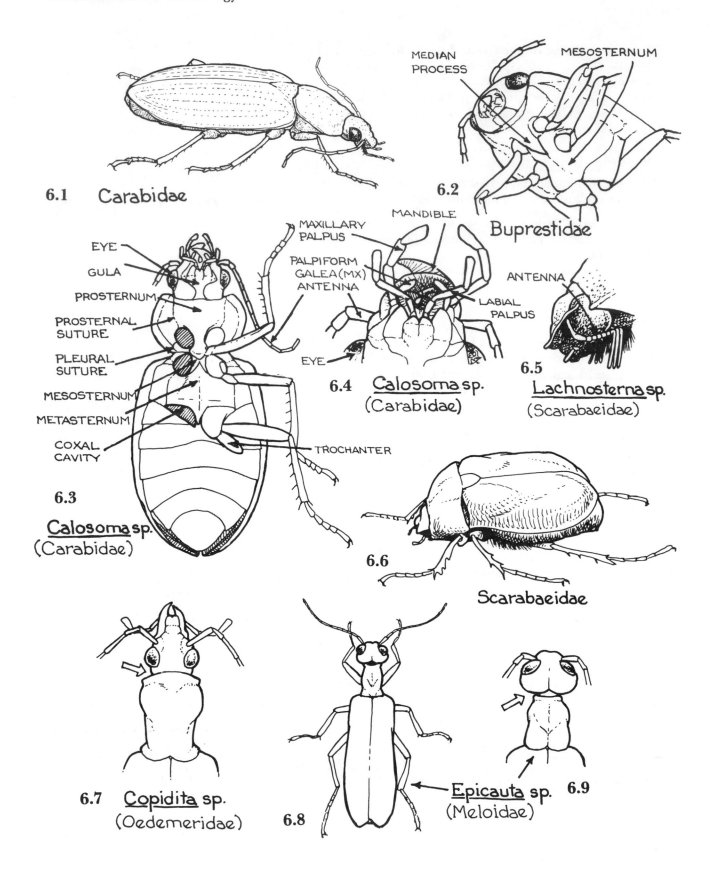

6.1   Carabidae

MEDIAN PROCESS

MESOSTERNUM

6.2   Buprestidae

EYE
GULA
PROSTERNUM
PROSTERNAL SUTURE
PLEURAL SUTURE
MESOSTERNUM
METASTERNUM
COXAL CAVITY

TROCHANTER

6.3
Calosoma sp.
(Carabidae)

MAXILLARY PALPUS
MANDIBLE
PALPIFORM
GALEA (MX)
ANTENNA
LABIAL PALPUS
EYE

6.4   Calosoma sp.
(Carabidae)

ANTENNA

6.5   Lachnosterna sp.
(Scarabaeidae)

6.6   Scarabaeidae

6.7   Copidita sp.
(Oedemeridae)

6.8

Epicauta sp.   6.9
(Meloidae)

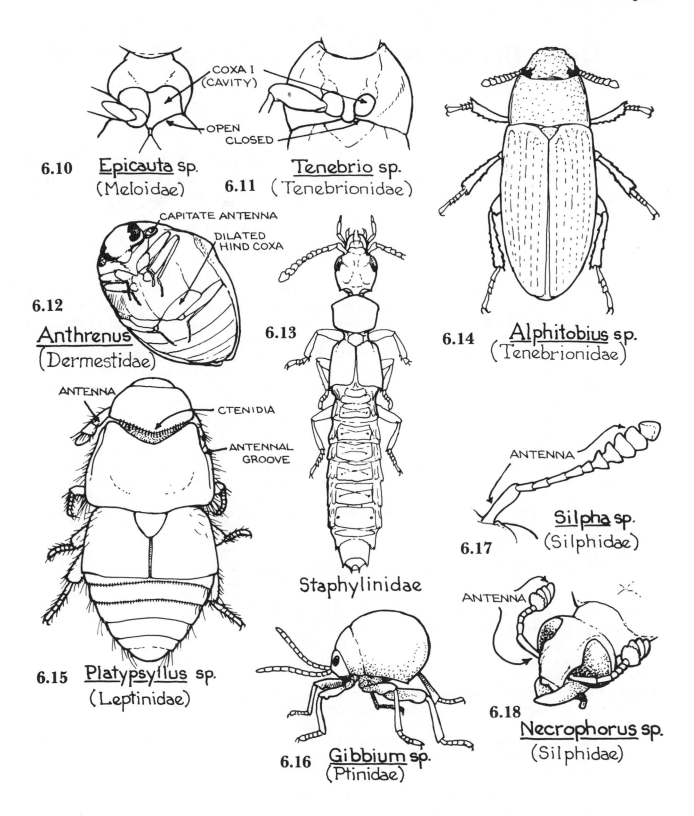

6.10 <u>Epicauta</u> sp.
(Meloidae)

COXA I (CAVITY)

OPEN
CLOSED

6.11 <u>Tenebrio</u> sp.
(Tenebrionidae)

CAPITATE ANTENNA

DILATED HIND COXA

6.12 <u>Anthrenus</u>
(Dermestidae)

6.13 Staphylinidae

6.14 <u>Alphitobius</u> sp.
(Tenebrionidae)

ANTENNA

CTENIDIA

ANTENNAL GROOVE

6.15 <u>Platypsyllus</u> sp.
(Leptinidae)

ANTENNA

6.17 <u>Silpha</u> sp.
(Silphidae)

ANTENNA

6.18 <u>Necrophorus</u> sp.
(Silphidae)

6.16 <u>Gibbium</u> sp.
(Ptinidae)

7   Order Hemiptera

The Hemiptera, or true bugs, consists of two major subdivisions: Homoptera with about 32,000 species and Heteroptera with about 23,000 species. Members of this order have incomplete metamorphosis (five nymphal instars to adult), piercing-sucking mouthparts (see Exercise 2), and two pairs of wings. In the Homoptera the wings are uniformly thickened, whereas in the Heteroptera the apical half of the forewing is membranous in contrast with the thickened basal half. The Homoptera includes the aphids, leaf- and treehoppers, spittle bugs, cicadas, and scale insects; they are of little importance to medical entomology. The Heteroptera includes about 50 families of plant bugs, water bugs, assassin bugs, and stink bugs. Only three families include obligate blood-sucking species: Reduviidae (assassin bugs), Cimicidae (bed bugs), and Polyctenidae (bat bugs). In the family Reduviidae only the subfamily Triatominae, known as chinches de pincudos, kissing bugs, or cone nose bugs, includes obligate blood-sucking species, but others of the family are of minor importance because of their painful bites. The three groups – kissing bugs, bed bugs, and bat bugs – represent a progressive sequence of morphological adaptations for ectoparasitism. Kissing bugs show few adaptations different from their predaceous relatives. Bed bugs possess a number of modifications, differing from their near relatives Anthocoridae, that better adapt them to close association with a blood source. In contrast, bat bugs have numerous morphological traits associated with a much higher degree of ectoparasitism than their relatives, the Cimicidae (Usinger, 1966).

With the aid of the following keys, identify the specimens of Hemiptera available. A number of terms and structures are peculiar to these groups. One deserves special note: In the Cimicidae, the spermalege (actually the ectospermalege of Usinger, 1966) relates to reproduction by traumatic insemination where the male punctures the body wall of the female during copulation (Usinger, 1966). Only females have the spermalege (males and females can be distinguished by caudal differences shown in Figures 7.21 and 7.22). Traumatic insemination is not a unique adaptation of the bed bugs; it also occurs in the bat bugs (Polyctenidae) and in two other related families. Other terms may be identified by study of the labeled illustrations in this exercise.

Key to principal hemipteran families of medical importance, adults only[a]

---

1. Base of rostrum usually not touching anterior coxae; forewings usually thickened basally and membranous apically, hindwings entirely membranous, both pairs folded flat over back with apices overlapping (Figure 7.1); legs that may be modified for swimming; tarsi normally 3-segmented; gular region usually well developed .... suborder Heteroptera 2
- Base of rostrum usually touching anterior coxae; wings usually of same texture throughout and folded rooflike over the back; legs never modified for swimming; tarsi 1- to 3-segmented; gular region absent, short, or membranous; plant feeders; some may prick the skin of humans occasionally (leafhoppers, spittle bugs, aphids, scales) .... suborder Homoptera

2. Antennae shorter than head, concealed in grooves beneath the eyes and invisible from above (Figure 7.16) .... series Cryptocerata 3
- Antennae longer than head and usually plainly visible from above, rarely hidden in grooves (Figure 7.1, 7.24) .... series Gymnocerata 6

3. Base of head overlapping front margin of pronotum; rostrum very short, scarcely distinguished from head; anterior tarsi 1-segmented, spatulate, fringed with short bristles; water bugs, natural enemies of mosquito larvae (water boatmen) .... Corixidae

- Base of head inserted in pronotum; rostrum cylindrical or cone-shaped and clearly segmented; anterior tarsi not as above 4

4. Membrane of hemelytra without distinct veins; abdomen without caudal appendages; front femora not greatly enlarged; predaceous water bugs which may bite humans painfully (back swimmers).....................................Notonectidae
   - Membrane of hemelytra with distinct veins; abdomen with a pair of short, straplike, or long, slender posterior appendages; front femora may be enlarged.......................................5

5. Hind legs flattened, fitted for swimming; caudal appendages short, sometimes concealed beneath tips of hemelytra. Predaceous water bugs; painful bite (giant water bugs, toe biters) ....Belostomatidae
   - Hind legs cylindrical, formed for walking; caudal appendages long and slender. Predaceous water bugs; severe biters (water scorpions) ....Nepidae

6. Claws of fore tarsi inserted before apex (Figure 7.14); tip of last tarsal segment more or less cleft 7
   - All claws inserted at tips of tarsi; tarsal apices entire 8

7. Hind femora very long, greatly exceeding apex of abdomen; predaceous bugs, of significance in natural control of mosquito larvae, etc. (water striders) ....Gerridae
   - Hind femora scarcely, if at all, exceeding tip of abdomen; predaceous on mosquito larvae and other small aquatic organisms (broad-shouldered water striders)......................Veliidae

8. Head as long as entire thorax; body linear; aquatic; predaceous on small aquatic organisms (water measurers, marshtreaders) ....Hydrometridae
   - Head seldom exceeding the combined length of pronotum and scutellum; body rarely very narrow; terrestrial 9

9. Rostrum appearing 3-segmented because the first segment is reduced, very short, or apparently absent (Figure 7.9) 10
   - Rostrum 4-segmented, the first segment well developed (Figure 7.15); phytophagous and some predaceous forms, including reported blood-suckers (*Clerada*) ....Lygaeidae

10. Head cylindrical; prosternum with a median longitudinal stridulatory groove between the fore coxae which receives the tip of the rostrum (lacking in *Linshcosteus*) (Figures 7.9, 7.10) (assassin bugs, cone noses) ....Reduviidae 14
    - Head rarely cylindrical; prosternum without a median longitudinal stridulatory groove between the fore coxae 11

11. Ocelli absent; clypeus broadened apically (Figure 7.24); hemelytra always greatly reduced, without membrane; small, blood-sucking parasites of warm-blooded animals 12
    - Ocelli present; clypeus with parallel or subparallel sides; hemelytra not reduced; predaceous on insects 13

12. Ctenidia present on head and antennae and often on pronotum and hemelytra. Hemelytra broadly attached to thorax, immovable. Bat ectoparasites (Figure 7.29) ....Polyctenidae

- Ctenidia absent. Hemelytra articulated at base, movable (bed bugs) (Figure 7.24) ....Cimicidae 40

13. Hemelytra with cuneus (Figure 7.17); membrane without long closed cells; meso–meta pleura divided; sometimes brachypterous; predators often found in bird and rodent nests; some will bite man (minute pirate bugs) ....Anthocoridae
    - Hemelytra without a cuneus; membrane with 4 or 5 closed cells (Figure 7.18); meso–meta pleura simple; predators feeding on shoreline (shore bugs) ....Saldidae

Reduviidae: key to some medically important subfamilies and species of Reduviidae and to tribes and genera of Triatominae of the world, adults[b]

14. Hemelytra lacking quadrangular or discoidal cell at base of membrane (Figure 7.2) 15
    - Hemelytra with a quadrangular or discoidal cell at the base of the membrane (Figure 7.3) ....Harpactorinae 38

15. Pronotum with only a slight constriction near the middle (Figure 7.2); front coxae not flattened, their outer sides convex 16
    - Pronotum with a pronounced transverse constriction behind or near the middle (Figures 7.4, 7.5); front coxae with outer sides flat or concave ....Piratinae 37

16. In lateral view, antennae inserted on top of head between margins close to eyes; rostrum stout, distinctly curved, front of head turned downward (Figure 7.7) (a common predator in N. America; painful bite).....................Reduviinae, e.g., *Reduvius personatus*
    - In lateral view, antennae inserted in lateral or dorsolateral margins of head; rostrum slender and relatively straight (Figure 7.9); head strongly produced anteriorly (Figures 7.9, 7.10) (blood-sucking parasites)........................Triatominae 17

17. Ocelli inconspicuous, not elevated above level of general integument, located in or very near interocular sulcus 18
    - Ocelli conspicuous, on distinct elevations on disc of postocular region (Figures 7.8, 7.13) 22

18. Head ovoid, in profile strongly convex dorsally; genae not extending beyond level of apex of clypeus; ocelli in or close behind strongly curved interocular sulcus; corium with veins indistinct; body integument smooth (Panama, S. America)..............(Cavernicolini) *Cavernicola*
    - Head elongated, subconical, not strongly convex dorsally in profile; genae large, elongate, extending beyond apex of clypeus; ocelli on postocular region, interocular sulcus lacking; corium with distinct veins; body integument heavily rugose and granulose (Mexico, Central and S. America, Cuba) ....(Bolboderini) 19

19. Basal segment of rostrum as long as or longer than second segment; scutellum with basal, triangular, sublateral projections; dorsal connexivum with submarginal ridge ....*Belminus*
    - Basal segment of rostrum much shorter than second segment; scutellum without basal sublateral projections; dorsal connexivum plain 20

20. Scutellum trapezoidal, hind margin straight across, without posterior process...*Parabelminus*
    - Scutellum triangular, with well-developed posterior process 21

21. Genae compressed laterally; femora without spines; tarsi
2-segmented, shorter than one-fifth length of tibia . . . .*Microtriatoma*
   - Genae spiniform; femora with spines; tarsi 3-segmented, about
one-third as long as tibia . . . .*Bolbodera*

22. Head with distinct lateral callosities immediately behind the
eyes (Figure 7.6); antennae inserted close to apex of head
(southern Mexico to Argentina) . . . .(Rhodniini) 23
   - Head without such callosities; antennal insertion remote from
apex of head (Figures 7.9–7.13) 24

23. Head subtriangular, dorsally flattened, less than twice as long
as wide at level of eyes; postocular region very short; apical
segment of rostrum deeply emarginated distally; femora
widened and laterally compressed . . . .*Psammolestes*
   - Head subcylindrical, not flattened dorsally, at least twice as
long as wide at level of eyes (Figure 7.6); postocular region
long, at most twice as wide as long; last segment of rostrum
pointed distally; femora of most elongate, subcylindrical, not
compressed . . . .*Rhodnius* (e.g., *R. prolixus*)

24. Small species, approximately 5 mm in length; head very short
and wide, length and width subequal (Venezuela) . . . .(Alberproseniini) *Alberprosenia*
   - Size greater than 5 mm in length; head elongate, longer than
wide at eye level (northern U.S. to southern Chile and
Argentina; China to northern Australia; Indian Ocean
and Africa). . . . .(Triatomini) 25

25. Head short and wide; antenniferous tubercles close to anterior
margin of eyes (Figure 7.8); head and body glabrose or with
short appressed setae . . . .*Panstrongylus*
   - Head variable, usually subcylindrical; antenniferous tubercles
not close to eyes 26

26. Rostrum short, not extending back beyond level of eyes
(Figure 7.13); prosternum without stridulatory groove . . . .*Linshcosteus*
   - Rostrum extending back to prosternum which has stridulatory
groove (Figure 7.10) 27

27. Posterior process of scutellum as long, oblique, sharply pointed
spine (as long as scutellum); basal segment of rostrum
subequal in length to second segment (Figure 7.10) . . . .*Eratyrus*
   - Posterior process of scutellum not as above; basal segment of
rostrum distinctly shorter than second segment 28

28. Head, body, and appendages with abundant long, curved
hairs; head strongly convex above; eyes small (Figure 7.11);
length 12–15 mm . . . .*Paratriatoma*
   - Head and body appearing glabrose or with short, sparse hairs;
head not strongly convex above; eyes larger (Figure 7.9); body
length 9–42 mm 29

29. Very large (33–42 mm in length); femora not denticulate;
ventral connexival sclerites not visible (restricted to Baja,
California) . . . .*Dipetalogaster* e.g., *D. maximus*
   - Length rarely more than 33 mm; femora may or may not be
denticulate; ventral connexival sclerites distinct (Figure 7.2) . . . .*Triatoma* 30

30. Length greater than 24 mm 31
   - Length less than 24 mm 32

31. Connexival segments each transversly marked with yellow
    along posterior third; coria irregularly and narrowly pale
    basally and subapically (Texas and Mexico) ....*T. gerstaeckeri*
    - Connexivum pale longitudinally along lateral margins; coria
      nearly unicolorous or scarcely pale along elevated veins
      (Arizona and Mexico) ....*T. recurva*

32. Upper surface distinctly hairy, the hairs rather short, stiff, and
    dense and rostrum hairy (southeastern U.S., Texas, New
    Mexico) ....*T. lecticularius*
    - Upper surface entirely naked or with very sparse,
      inconspicuous, short appressed hairs; rostrum scarcely to
      densely hairy 33

33. First antennal segment reaching or almost reaching apex of
    clypeus 34
    - First antennal segment distinctly shorter, not reaching apex of
      clypeus, or, if approaching this, with scutellum turned down at
      apex and eyes small 35

34. Antero-lateral pronotal spines strongly, obliquely produced,
    subrounded at apices; scutellum shorter than broad,
    subtriangular. Tropicopolitan (Florida) ....*T. rubrofasciata*
    - Antero-lateral pronotal spines very short and rounded;
      scutellum about as long as broad, its posterior prolongation
      subcylindrical near apex (southwestern U.S., Mexico) ....*T. rubida*

35. Eyes relatively large, about one-half or more as wide as width
    of interocular space (Figure 7.12); posterior prolongation of
    scutellum long, cylindrical, not turned downward at apex;
    rostrum slender and scarcely hairy; antero-lateral tubercles of
    pronotum distinct (southeastern and southwestern U.S.,
    Mexico) ....*T. sanguisuga*
    - Eyes smaller, always less than one-half the width of interocular
      space; posterior prolongation of scutellum short, tapering
      throughout, and turned downward at apex; rostrum stout,
      densely hairy at apex; antero-lateral tubercles of pronotum
      vestigial 36

36. Body surface highly polished, shining; venter abruptly
    flattened; connexivum conspicuously alternated with yellow
    ochraceous and black, the middle of each segment with a
    broad black mark extending to lateral margin (Texas) ....*T. neotomae*
    - Body surface dull or moderately polished; venter rounded;
      connexivum entirely dark brown, brown with pale lateral
      margins, or alternated with light brown and reddish
      ochraceous (southwestern U.S. and Mexico) ....*T. protracta*

37. Hemelytra entirely black; apical portion of anterior tibiae
    angulately dilated beneath, the spongy fossa being preceded
    by a small prominence. Predaceous, painful biter (eastern
    U.S.) ....*Melanolestes picipes*
    - Corium and membrane of hemelytra each marked with a
      yellow spot; tibiae not dilated, the spongy fossae elongate.
      Predaceous, painful biter. (California and western U.S.) ....*Rasahus thoracicus*
      (southern U.S.) ....*Rasahus biguttatus*

38. Basal segment of beak about as long as front portion of head anterior to eyes; sides of mesosternum without a tubercle or fold. Predaceous bugs which may bite painfully ....*Zelus* spp.
    - Basal segment of beak longer than front portion of head; sides of mesosternum with tubercle or fold in front of the hind angles of the prosternum 39

39. Pronotum produced posteriorly over scutellum and with a high, median, tuberculate ridge (Figure 7.19); robust. Predaceous, painful biter (southern and eastern U.S.) (wheel bug) ....*Arilus cristatus*
    - Pronotum not produced or elevated as above; slender. Predaceous, painful biter (western U.S.) (spined soldier bug) ....*Sinea diadema*

Cimicidae: key to some representative genera and species[c]

40. Reduced hemelytra circular in outline (Figure 7.23); tibia mottled; first and second antennal segments subequal in length; large size (7 mm or more in length). On bats in Western Hemisphere ....*Primacimex* (Texas and Guatemala)..(*P. cavernis*)
    - Reduced hemelytra rectangular in form, broader than long (Figure 7.24); tibia not mottled; first antennal segment much shorter than second segment; size less than 7 mm in length 41

41. Bristles at sides of pronotum minutely serrate on outer edges (Figure 7.25); metasternum usually a flat spatulate plate; spermalege ventral on anterior margin of fifth visible segment (Figure 7.24)...42
    - Bristles at sides of pronotum serrate or cleft only at tips (Figure 7.28); metasternum reduced or convex; spermalege dorsal, lateral, or lacking 47

42. Metasternum partly compressed by coxae; spermalege in middle of abdominal segment. On bats in neotropics ....*Propicimex*
    - Metasternum a broad, discrete intercoxal plate; spermalege located right of midline...43

43. Third and fourth antennal segments subequal in length; length of second antennal segment two-thirds or less than width of interocular space; body pubescence long, pale, silk hairs. On swallows of Holarctic ....*Oeciacus* (American swallow bug, *O, vicarius*)
    - Fourth antennal segment only two-thirds of third in length (Figure 7.24); length of second antennal segment subequal to width of interocular space; body pubescence not as above ....*Cimex* 44

44. Width of pronotum two times the length at midline (Figure 7.24). On bats, chickens, and humans; tropicopolitan (tropical bed bug) ....*C. hemipterus*
    - Width of pronotum three times the length at midline...45

45. Hairs of pronotal fringe not longer than width of eye; contiguous portion of hemelytra about one-half the length of scutellum. On bats, chickens, and humans; cosmopolitan (common bed bug) ....*C. lectularius*
    - Hairs of pronotal fringe longer than width of eye; contiguous portion of hemelytra at least subequal to length of scutellum. On bats (pilosellus complex, 6 spp.)...46

46. Bristles of pronotal fringe with few serrations on outer edges. On bats in eastern U.S. to Colorado ....*C. adjunctus*
 • Bristles of pronotal fringe heavily serrated on outer edges. On bats in western U.S. ....*C. pilosellus*

47. No visible spermalege; third antennal segment longer than others combined; tibia without apical tufts. On bats and humans in Africa ....*Leptocimex (L. boueti)*
 • Spermalege present; third antennal segment shorter than others combined; tibia with apical tufts 48

48. Bristles at sides of pronotum numerous and subequal in size; spermalege at right ventrolateral anterior margin of sixth visible abdominal segment (Figure 7.27). Associated with woodpecker cavities in western U.S. and Mexico ....*Hesperocimex* 49
 • One or two conspicuously longer bristles at posterolateral angle of pronotum; spermalege dorsal (Figure 7.30) 51

49. Pronotum wide, 3 times as wide as long at midline (Figure 7.26); hemelytra pale in color (woodpecker nest cavities and man-made bird houses) ....*H. coloradensis*
 • Pronotum not more than 2.5 times as wide as long; hemelytra dark in part or whole 50

50. Hemelytra uniformly dark in color (woodpecker nest cavities) ....*H. sonorensis*
 • Hemelytra dark on edges, pale in middle (woodpecker nest cavities) ....*H. cochimiensis*

51. Head length and width subequal; spermalege right dorsolateral on edge of fifth segment. On white-throated swift in Nebraska and California ....*Synxenoderus comosus*
 • Head wider than long; spermalege at middle or right of midline 52

52. Rostrum long, reaching middle coxal base; spermalege at middle anterior margin of fifth segment. On bird of prey and poultry in southwestern U.S. and Mexico (Mexican chicken bug) ....*Haematosiphon inodorus*
 • Rostrum short, not reaching procoxal base; spermalege to right of middle 53

53. Pronotum 2.5 or more times wider than long. Associated with poultry (Brazil, Florida, Georgia) ....*Ornithocoris O. toledoi* and *O. pallidus)*
 • Pronotum about twice as wide as long; on chimney swifts in eastern N. America (Chimney Swift bug) ....*Cimexopsis nyctalis*

[a]Prepared with the advice of R. L. Usinger.
[b]Modified from Lent and Wygodzinsky (1979).
[c]Modified from Usinger (1966).

References

Fracker, S. B., and R. L. Usinger, 1949. Generic identification of Nearctic reduviid nymphs (Hemiptera). *Ann. Entomol. Soc. Am. 42*:273 – 8.

Lent, H., and P. Wygodzinsky. 1979. Revision of the Triatominae (Hemiptera, Reduviidae) and their significance as vectors of Chagas disease. *Bull. Am. Mus. Nat. Hist. 163*:125–520.

Miller, N. C. E. 1971. *The Biology of the Heteroptera*, 2nd ed. E. W. Classey, Hampton Middlesex, England. 106 pp.

Ueshima, N. 1968. New species and records of Cimicidae with keys. *Pan. Pac. Entomol. 44*:264–79.

Usinger, R. L. 1934. Blood-sucking among phytophagous Hemiptera. *Canad. Entomol. 66*:97–100.

– 1966. *Monograph of Cimicidae*, vol. 2. Thomas Say Foundation, Entomological Society of America, College Park, Md.

Usinger, R. L., and N. Ueshima. 1965. New species of bat bugs of the *Cimex pilosellus* complex. *Pan. Pac. Entomol. 41*:114–17.

Usinger, R. L., P. Wygodzinsky, and R. E. Ryckman. 1966. The biosystematics of Triatominae. *Ann. Rev. Entomol. 11*:309–30.

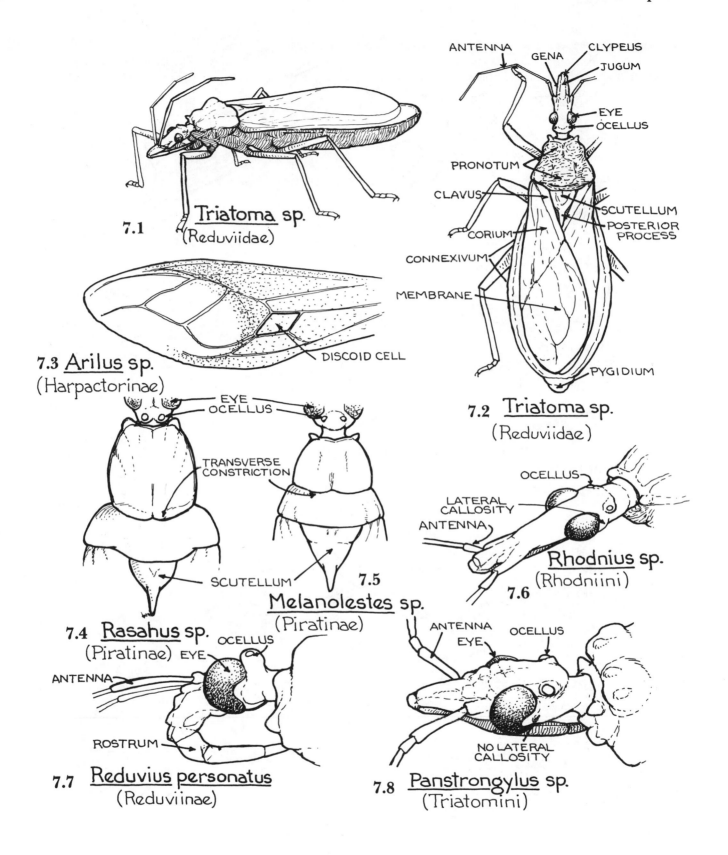

7.1 Triatoma sp.
(Reduviidae)

7.2 Triatoma sp.
(Reduviidae)

ANTENNA GENA CLYPEUS JUGUM EYE OCELLUS PRONOTUM CLAVUS SCUTELLUM POSTERIOR PROCESS CORIUM CONNEXIVUM MEMBRANE PYGIDIUM

7.3 Arilus sp.
(Harpactorinae)

DISCOID CELL

EYE OCELLUS TRANSVERSE CONSTRICTION SCUTELLUM

7.4 Rasahus sp.
(Piratinae)

7.5 Melanolestes sp.
(Piratinae)

7.6 Rhodnius sp.
(Rhodniini)

OCELLUS LATERAL CALLOSITY ANTENNA

7.7 Reduvius personatus
(Reduviinae)

OCELLUS EYE ANTENNA ROSTRUM

7.8 Panstrongylus sp.
(Triatomini)

ANTENNA EYE OCELLUS NO LATERAL CALLOSITY

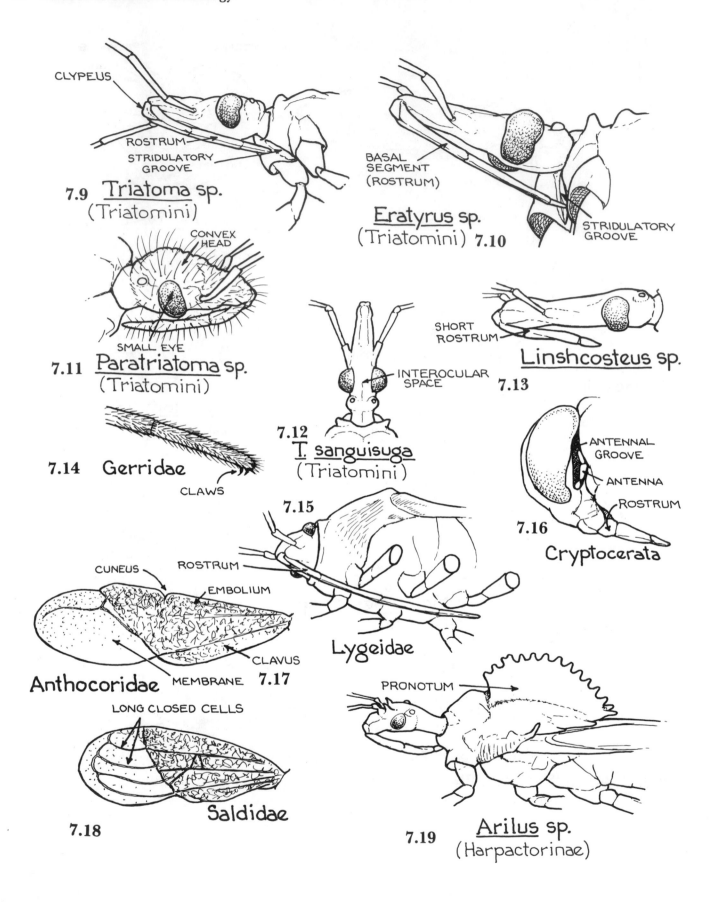

CLYPEUS

ROSTRUM

STRIDULATORY GROOVE

7.9 Triatoma sp. (Triatomini)

BASAL SEGMENT (ROSTRUM)

STRIDULATORY GROOVE

Eratyrus sp. (Triatomini) **7.10**

CONVEX HEAD

SMALL EYE

7.11 Paratriatoma sp. (Triatomini)

INTEROCULAR SPACE

7.12 T. sanguisuga (Triatomini)

SHORT ROSTRUM

Linshcosteus sp.

7.13

7.14 Gerridae

CLAWS

ANTENNAL GROOVE

ANTENNA

ROSTRUM

7.16 Cryptocerata

CUNEUS

ROSTRUM

EMBOLIUM

7.15

CLAVUS

Anthocoridae

MEMBRANE   **7.17**

Lygeidae

LONG CLOSED CELLS

Saldidae

7.18

PRONOTUM

7.19 Arilus sp. (Harpactorinae)

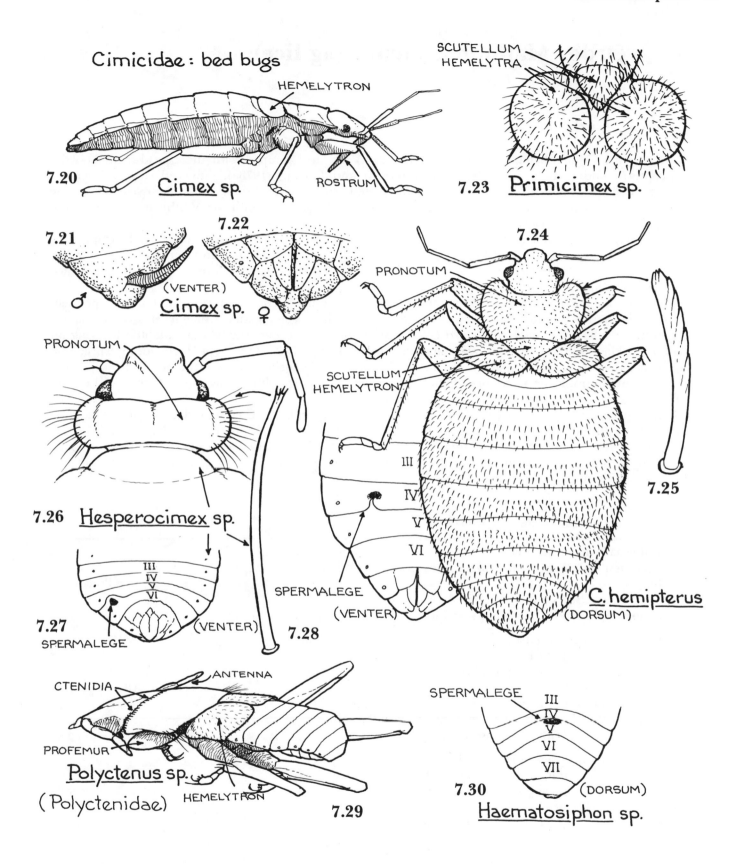

Cimicidae: bed bugs

7.20 **Cimex** sp.
HEMELYTRON
ROSTRUM

7.23 **Primicimex** sp.
SCUTELLUM
HEMELYTRA

7.21 ♂
7.22 (VENTER)
Cimex sp. ♀

7.24
PRONOTUM

7.26 **Hesperocimex** sp.
PRONOTUM

SCUTELLUM
HEMELYTRON

7.25

III
IV
V
VI

SPERMALEGE
(VENTER)

C. hemipterus
(DORSUM)

7.27
SPERMALEGE
III
IV
V
VI
(VENTER)
7.28

Polyctenus sp.
(Polyctenidae)
CTENIDIA
ANTENNA
PROFEMUR
HEMELYTRON
7.29

7.30
SPERMALEGE
III
IV
V
VI
VII
(DORSUM)
Haematosiphon sp.

8   Order Mallophaga (chewing lice)

Both chewing (Mallophaga) and sucking (Anoplura) lice are grouped with the psocids and booklice (Psocoptera) under the superorder Psocodea (Kim and Ludwig, 1978a), thus reflecting a common phylogenetic origin that probably occurred during the Permian period (some 250 million years ago). The order Mallophaga includes three suborders (Amblycera, Ischnocera, and Rhynchophthirina), representing nearly 3,000 known species.

Mallophaga spend their entire lives on a host. Their eggs are attached to host pelage, and nymphs and adults have a size, habitus, and behavior that aid in maintaining continual host contact. Hosts become infested by contact with each other or sometimes through other ectoparasites (paratenic hosts), to which small lice may attach.

The mandibulate adult lice range in length from 1 to 11 mm. Their body form is flattened (depressed) (Figures 8.4 and 8.5), often heavily sclerotized, and may have color patterns of darker spots or bands. Apparently Mallophaga evolved as bird ectoparasites and have secondarily become adapted to living on mammals. They characteristically have a high degree of host specificity. Present phylogenetic interpretation holds that the two mallophagan suborders, Ischnocera and Rhynchophthirina (the elephant lice), arose from the protoanopluran ancestors of the sucking lice (order Anoplura, Exercise 9) (Kim and Ludwig, 1978a).

Louse specimens may be collected directly from dead or living hosts. Collection data should include the host species and site on the host. Lice can be mounted permanently on slides or kept free in liquid preservative for keying and identification. Keys to the orders of Psocodea and to all families of Mallophaga found in North America are given below, along with keys to common species found on some mammals and those occurring on the domestic pigeon and chicken.

Key to orders and suborders of Psocodea[a]

---

1. Antennae with more than 10 segments; hypognathous head; abdomen with 7 or 8 pairs of spiracles (psocids and book lice) .... order Psocoptera
   - Antennae with 5 or fewer segments; head prognathous; abdomen with 6 or fewer pairs of spiracles 2

2. Piercing-sucking mouthparts that withdraw into an internal, trophic sac; notum of thorax greatly reduced; sitophore (esophageal sclerite) lacking; thorax wider than head (sucking lice) (Exercise 9) .... order Anoplura
   - Chewing mouthparts, mandibles sclerotized; notum of thorax well developed; sitophore present; head wider than thorax or prolonged into a long proboscis (chewing lice) .... order Mallophaga 3

3. Maxillary palpi present (Figure 8.1, 8.2); antenna usually 4-segmented, capitate or clavate (may be completely or partly concealed in ventral grooves) .... suborder Amblycera 5
   - Maxillary palpi absent (Figure 8.3); antenna 3- or 5-segmented, usually filiform and exposed 4

4. Pronotum strongly developed; thoracic segmentation distinct; forehead not developed into a long proboscis .... suborder Ischnocera 10
   - Pronotum much reduced; thoracic segmentation indistinct; forehead developed into a long prognathous proboscis with terminal mandibles (Figure 8.8) (on elephants and wart hogs) .... suborder Rhynchophthirina

Key to families of Mallophaga found in America north of Mexico, adults only[b]

Families of Amblycera

5. All tarsi with 2 claws 6
 • Tarsi of mid- and hindlegs with one or no claw (Figure 8.6)
   (7 genera; on neotropical rodents) ....Gyropidae

  Gyropidae

 • Tarsi with at least 1 claw each; 6 pairs of abdominal spiracles
   present; maxillary palpi 3- or 4-segmented ....*Gyropus*
   (oval guinea pig louse) *Gyropus ovalis*
 • Tarsi without claws (Figure 8.6); 5 pairs of abdominal
   spiracles; maxillary palpi 2-segmented ....*Gliricola*
   (slender guinea pig louse) *Gliri-
   cola porcelli*

6. Antennae clavate (Figure 8.10) (gradual thickening into club
   shape), 4- or 5-segmented (if 5 then terminal segments are not
   compact) 7
 • Antennae capitate (abruptly thickened into compact terminal
   club); body with slender spinelike setae (8 genera; on dogs,
   kangaroos, wallabies, and cassowary) (on dogs: *Heterodoxus
   spiniger*) ....Boopidae

7. Six pairs of abdominal spiracles; prothorax and mesothorax
   not fused (Figure 8.1) 8
 • Only 5 abdominal segments with spiracles; mesothorax either
   greatly reduced or fused with prothorax (6 genera; on Central
   and S. American rodents and marsupials) (on guinea pig:
   Trimenopon hispidum) ....Trimenoponidae

8. Head not broadly triangular, not evenly expanded behind or
   strongly enlarged at temples. Antennal capsule lateral 9
 • Head broadly triangular, strongly enlarged at temples and
   evenly enlarged behind. Antennal capsule ventral (Figures 8.1,
   8.2) (60 genera in America north of Mexico, on poultry and
   other birds) ....Menoponidae
   (see key to lice of chickens)

9. Antennal capsules as pronounced lateral swellings in front of
   eyes (Figure 8.7); abdomen without lateral intersegmental
   notching (1 genus, *Laemobothrion;* on birds) ....Laemobothriidae
 • Antennal capsules not pronounced; sides of head straight or
   concave (Figures 8.9, 8.10); abdomen sometimes notched
   laterally (3 genera, *Ricinus, Trochiloecetes, Trochiliphagus*) ....Ricinidae

Families of Ischnocera

10. Abdomen consisting of more than 7 segments (Figure 8.4) 11
 • Abdomen with only 7 visible segments (21 genera; on birds,
   Tinamous, of Central and S. America) (Figure 8.12) ....Heptapsogasteridae

11. Antennae with 3 segments in male and usually in female; tarsi
   with 1 claw (Figure 8.11) (14 genera; on mammals) ....Trichodectidae 12
 • Antennae with 5 segments in both sexes; tarsi with 2 claws
   (Figure 8.5) (130 genera; on birds) ....Philopteridae
   (see key to lice of chickens or pigeons)

Trichodectidae

12. Three pairs of abdominal spiracles (Figure 8.16); second
    antennal segment of male longer than base segment 13
  - Six pairs of abdominal spiracles or none evident; second
    antennal segment of male shorter than base segment 14

13. Preantennal region triangular (Figure 8.16) (on cats) . . . .*Felicola subrostrata*
  - Preantennal region broadly rounded (on foxes). . . . .*Suricatoecus vulpis*

14. Head width equal to or less than length 15
  - Head width greater than length (Figure 8.17) 23

15. Preantennal region broadly rounded with numerous short
    setae (if notched anteriorly, then male and female antennae
    similar); abdominal segments each with a transverse row of
    setae and with scattered setae (on cows, goats, horses, sheep,
    elk, and reindeer) . . . .*Bovicola* (= *Damalinia*) 16
  - Preantennal region with anterior margin set off by distinct
    angles from lateral margins, with sparse short setae; sexes with
    dissimilar antennae; abdominal segments each with a
    transverse row of setae 21

16. Body sparsely covered with short setae 17
  - Body densely covered with long setae (from Angora goat) . . . .*B. crassipes*

17. Antennae of sexes dissimilar; first segment of male antennae
    considerably swollen 18
  - Antennae of sexes similar 19

18. Gonapophysis of female with internal margin nearly straight
    and with internal lobe reduced (Figure 8.14) (on horses) . . . .*B. equi*
  - Gonapophysis of female with large internal lobe or projection
    (on sheep) . . . .*B. ovis*

19. Anterior margin of female clypeus feebly emarginate
    (notched) or not at all 20
  - Anterior margin of female clypeus deeply emarginate (on
    Angora goats) . . . .*B. limbatus*

20. Head subquadrangular (Figure 8.13); dorsal surface of head
    with few setae; setae on inner margin of gonapophysis
    restricted to lobe (on goats) . . . .*B. caprae*
  - Head not quadrangular, narrowing anteriorly; dorsal surface
    of head with numerous setae; setae on inner margin of
    gonapophysis not restricted to lobe (on cattle) . . . .*B. bovis*

21. Male with parameres fused distally; endomeres fused to form
    endomeral plate (Figure 8.19) (pseudopenis) (on deer) . . . .*Tricholipeurus* 22
  - Parameres and endomeres free (Figure 8.20) (on
    porcupines) . . . .*Eutrichophilus*

22. First antennal segment less than half the total antennal length
    (on deer) . . . .*T. parallelus*
  - First antennal segment of male at least half the total antennal
    length (on deer) . . . .*T. lipeuroides*

23. Six pairs of abdominal spiracles present; antennae of male
    shorter than head 24
  - No visible abdominal spiracles; antennae of male longer than
    head (Figure 8.17) (on pocket gophers) . . . .*Geomydoecus*

24. Parameres and endomeres free (see Figure 8.20) 25
    - Parameres fused distally (see Figure 8.19); endomeres fused
      to form endomeral plate (on wild carnivores: badgers,
      cacomistles, skunks, and probably weasels) ....*Neotrichodectes* 7 spp.

25. Antennae of the sexes similar (on minks, martins, ermines, and
    weasels) ....*Stachiella* 4 spp.
    - Antennae sexually dimorphic (on dogs, wolves, bears, coyotes,
      and racoons) ....*Trichodectes* 3 spp.
      (on dogs and other canines) ....*Trichodectes canis*

[a]Modified from Kim and Ludwig (1978a).
[b]Prepared with aid of John S. Wiseman and Bernard C. Nelson.

Key to species of Mallophaga occurring on the domestic chicken[a]

1. Maxillary palpi present (suborder Amblycera,
   Menoponidae) 2
   - Maxillary palpi absent (suborder Ischnocera, Philopteridae) 5

2. Forehead armed ventrally with a pair of prominent spinelike
   processes 3
   - Forehead without ventral spinelike processes ....*Menopon gallinae*

3. Abdominal tergites III–VII each with two transverse rows of
   setae 4
   - Abdominal tergites III–VII each with one transverse row of
     setae ....*Menacanthus pallidulus*

4. More than 2 mm in length. Numerous short setae scattered on
   dorsum of meso-metathorax ....*Menacanthus stramineus*
   - Less than 2 mm in length. Dorsum of meso-metathorax with a
     few short setae on the lateral margins ....*Menacanthus cornutus*

5. Head longer than wide 8
   - Head wider than long 6

6. Two long setae on each side of the head in the postantennal
   region; remainder of the chaetotaxy of the head with only
   short setae (Figure 8.21) ....*Goniocotes gallinae*
   - More than two long setae on each side of the head in the
     postantennal region; additional long setae on the dorsum of
     the head 7

7. Antennae similar in the two sexes. Three long setae on each
   temple ....*Goniodes gigas*
   - Antennae sexually dimorphic. Two long setae on each
     temple ....*Goniodes dissimilis*

8. Terminal abdominal segment of male with a sternal process
   (Figure 8.15). Lateral margins of terminal abdominal segment
   of female each with a longitudinal row of setae ....*Oxylipeurus dentatus*
   - Terminal abdominal segment of male without a sternal
     process. Lateral margins of terminal abdominal segment of
     female each without a longitudinal row of setae 9

9. First antennal segment of male with an appendage (Figure
   8.3). Dorsally, 1 patch of long setae in each posterior lateral
   angle of pterothorax, with the intervening margin void of long
   setae 10

- First antennal segment of male without an appendage.
  Dorsally, 4 patches of long setae on posterior margin of
  pterothorax . . . .*Cuclotogaster heterographus*

10. Margin of vulva in female with a row of short setae (Figure
    8.18). In the male, postantennal constriction, greatest width of
    head in the preantennal region . . . .*Lipeurus caponis*
- Margin of vulva in female without a row of short setae. In the
  male, greatest width of head in the post-antennal region . . . .*Numidilipeurus tropicalis*

[a]From Emerson (1956).

Key to species of Mallophaga occurring on the Domestic Pigeon[a]

1. Maxillary palpi present (suborder Amblycera) 2
- Maxillary palpi absent (suborder Ischnocera) 4

2. Forehead armed ventrally with a pair of prominent spinelike
   processes . . . .*Hohorstiella lata*
- Forehead without ventral spinelike processes 3

3. Venter of third femora with a comb of fine setae . . . .*Colpocephalum turbinatum*
- Venter of third femora without a comb of fine setae . . . .*Bonomiella columbae*

4. Head longer than wide . . . .*Columbicola columbae*
- Head wider than long 5

5. Forehead armed ventrally with a pair of spinelike processes . . . .*Physconelloides zenaidurae*
- Forehead without ventral spinelike processes 6

6. Antennae filiform and similar in both sexes . . . .*Campanulotes bidentatus compar*
- Antennae sexually dimorphic; in the male, first segments
  enlarged and third segments elongated, and fourth and fifth
  segments greatly reduced . . . .*Coloceras damicorne fahrenholzi*

[a]From Emerson (1957).

References

Emerson, K. C. 1956. Mallophaga (chewing lice) occurring on the domestic chicken. *J. Kansas Entomol. Soc. 29*:63–79.

– 1957. A new species of Mallophaga from the pigeon. With a key to the species of Mallophaga occurring on the pigeon. *Florida Entomol. 40*:63–4.

Hopkins, G. H. E., and T. Clay. 1952. *A Checklist of the Genera and Species of Mallophaga*. British Museum of Natural History, London. 362 pp.

Kim, K. C., and H. W. Ludwig. 1978a. Phylogenetic relationships of parasitic Psocodea and taxonomic position of the Anoplura. *Ann. Entomol. Soc. Am. 71*:910–22.

– 1978b. The family classification of the Anoplura. *Syst. Entomol. 3*:249–84.

Malcomson, R. O. 1960. Mallophaga from birds of North America. *Wilson Bull. 72*:182–97.

Price, R. D. 1977. *Menacanthus* (Mallophaga, Menoponidae) of the Passeriformes (Aves). *J. Med. Entomol. 14*:207–20.

Scanlon, J. E. 1960. The Anoplura and Mallophaga of the Mammals of New York. *Wildlife Disease* No. 5 (3 microcards). Wildlife Disease Association, Ames, Iowa.

Werneck, F. L. 1948. *Os Malofagos de Mamiferos*, parte I. Edicao de Revista Brasileira de Biologia, Rio de Janeiro. 243 pp.

– 1950. *Os Malofagos de Mamiferos*, parte II. Edicao de Instituto Oswaldo Cruz, Rio de Janeiro. 207 pp.

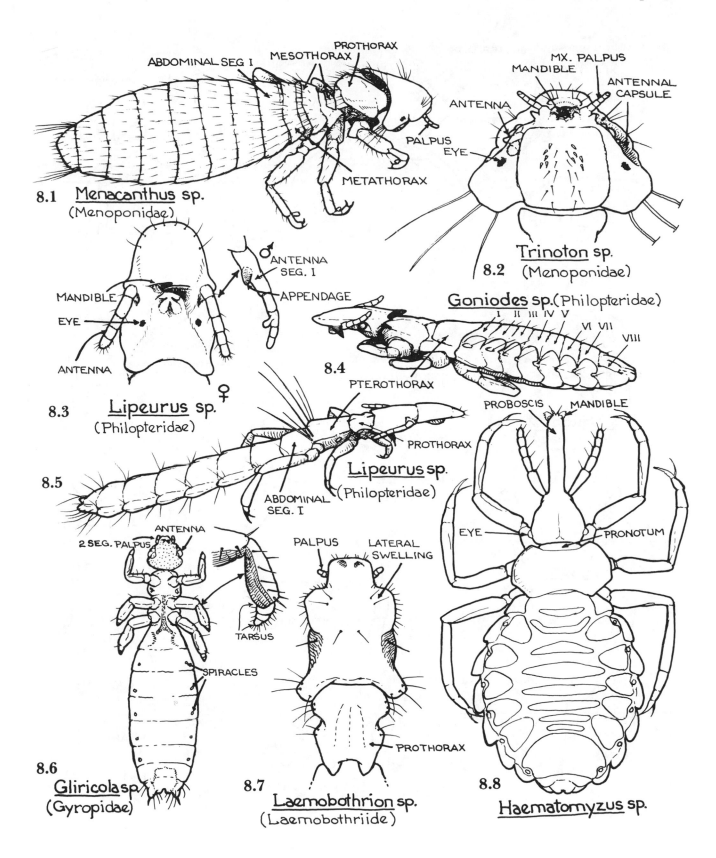

8.1 <u>Menacanthus</u> sp.
(Menoponidae)

ABDOMINAL SEG I
MESOTHORAX
PROTHORAX
METATHORAX
PALPUS

8.2 <u>Trinoton</u> sp.
(Menoponidae)

MANDIBLE
MX. PALPUS
ANTENNAL CAPSULE
ANTENNA
EYE

8.3 <u>Lipeurus</u> sp. ♀
(Philopteridae)

MANDIBLE
EYE
ANTENNA
ANTENNA SEG. I
APPENDAGE

<u>Goniodes</u> sp. (Philopteridae)

8.4

8.5 <u>Lipeurus</u> sp.
(Philopteridae)

PTEROTHORAX
PROTHORAX
ABDOMINAL SEG. I

2 SEG. PALPUS
ANTENNA
TARSUS
SPIRACLES

8.6 <u>Gliricola</u> sp.
(Gyropidae)

8.7 <u>Laemobothrion</u> sp.
(Laemobothriide)

PALPUS
LATERAL SWELLING
PROTHORAX

8.8 <u>Haematomyzus</u> sp.

PROBOSCIS
MANDIBLE
EYE
PRONOTUM

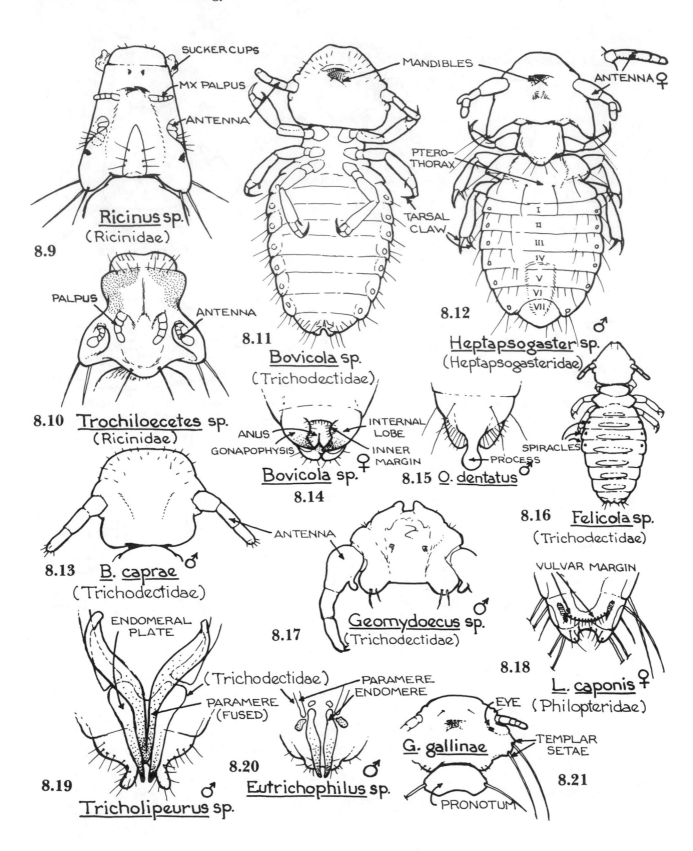

SUCKER CUPS
MX PALPUS
ANTENNA
8.9 <u>Ricinus</u> sp.
(Ricinidae)

PALPUS
ANTENNA
8.10 <u>Trochiloecetes</u> sp.
(Ricinidae)

MANDIBLES
PTERO-THORAX
TARSAL CLAW
8.11 <u>Bovicola</u> sp.
(Trichodectidae)

ANTENNA♀
I
II
III
IV
V
VI
VII
8.12 <u>Heptapsogaster</u> sp.
(Heptapsogasteridae)

ANUS
GONAPOPHYSIS
INTERNAL LOBE
INNER MARGIN
<u>Bovicola</u> sp. ♀
8.14

SPIRACLES
PROCESS
8.15 <u>O. dentatus</u> ♂

SPIRACLES
8.16 <u>Felicola</u> sp.
(Trichodectidae)

ANTENNA
8.13 <u>B. caprae</u> ♂
(Trichodectidae)

ANTENNA
<u>Geomydoecus</u> sp. ♂
8.17 (Trichodectidae)

VULVAR MARGIN
8.18 <u>L. caponis</u> ♀
(Philopteridae)

ENDOMERAL PLATE
(Trichodectidae)
PARAMERE (FUSED)
8.19 ♂
<u>Tricholipeurus</u> sp.

PARAMERE
ENDOMERE
8.20 ♂
<u>Eutrichophilus</u> sp.

EYE
<u>G. gallinae</u>
TEMPLAR SETAE
PRONOTUM
8.21

9 Order Anoplura (sucking lice)

Anoplura, meaning "unarmored tail," constitutes a small order of small wingless insects (1–6 mm in length) living exclusively as blood-sucking ectoparasites of mammals. At present there are about 500 known species classified into 15 families, but this represents only about one-half of the total estimated by Kim and Ludwig (1978). During the past 25 years there has been a 90% increase in the number of recognized species of sucking lice. Many families contain only one or a few genera, reflecting the known extinction of large groups of prehistoric mammalian hosts which are now poorly represented, but which once probably possessed a diverse complement of anopluran species that had coevolved with these hosts.

The more primitive evolutionary line of sucking lice is represented by Haematopinidae and by four other families including those on marine mammals (exclusive of cetaceans), humans, wild pigs, and camels. Linognathidae, Polyplacidae, and Hoplopleuridae represent the more advanced anoplurans, along with seven additional families; their hosts are mainly large ungulates, rodents, insectivores, and lagomorphs (Kim and Ludwig, 1978). There are no anoplurans known from a number of eutherian mammal orders: Monotremata, Marsupialia, Chiroptera, Edentata, Pholidota, Cetacea, Proboscidea, Sirenia, and most land Carnivora (Kim and Ludwig, 1978).

Sucking lice are highly modified for permanent ectoparasitism. All stages, including eggs, occur on the host. Characteristic adaptations include the retraction of biting stylets into a cephalic sac, a reduced notum which gives legs a more lateral placement, pretarsal development of a "thumb–claw" complex and the glueing of eggs to host pelage. These insects are highly host-specific, being very dependent upon the microenvironment provided by the host. Many are even restricted to specific sites on their host. This restriction may reflect either the microenvironment of the site or the host's self-grooming limitation for various parts of its body.

Infestation occurs through contact between hosts. Thus collection can be made directly from the host's skin or pelage. Collection data should include accurate identification of the host to species and a record of the site (or sites) occupied on the host. For identification, lice should be cleared and mounted on microscope slides; unmounted lice are stored in 70% alcohol.

Keys to the families of Anoplura are given below, along with those of some important genera and species.

Key to families of Anoplura[a]

1. Head with distinct eyes (Figure 9.2) or subacute ocular points
   (Figure 9.3) on postantennal margins 2
   ● Head without eyes or subacute ocular points (Figure 9.4)..8

2. Head with ocular points but without eyes (Figure 9.3) 3
   ● Head with eyes (having distinct lens) but without ocular
   points 4

3. Thorax with sternal plate well developed (Figure 9.7);
   paratergal plates as sclerotized caps, without margins free
   from abdomen (Figure 9.19); legs subequal in size (1 genus, 22
   spp. on ungulates) ....Haematopinidae *Haematopinus* 15
   ● Thorax without sternal plate; paratergal plates with posterior
   margins free from abdomen; forelegs much smaller than mid-
   and hindlegs (1 genus, 1 sp. on Tubulidentata: aardvark) ....Hybophthiridae

4. Abdomen sparsely covered with setae, paratergal plates as
   tubercles, caps, or lobes; head attached to anterior end of
   thorax 5

- Abdomen densely covered with fine setae, paratergal plates lacking; head attached to thorax dorsally (1 genus, 4 spp. on Camelidae) .... Microthoraciidae

5. Head much longer than thorax (Figure 9.8); abdomen elliptical in outline, with tuberculate paratergal plates (Figure 9.8) (1 genus, 1 sp., *Pecaroecus javalii,* on peccaries, Tayassuidae) ....Pecaroecidae
- Head and thorax subequal in length; abdomen oval or elliptical, paratergal plates as sclerotized caps or lobes 6

6. Body less than twice as long as wide; thorax very wide; abdomen oval, as broad as thorax basally, with prominent tuberculate lateral lobes; forelegs slender (Figure 9.17) (1 genus, 2 spp., on Hominidae and Pongidae) ....Pthiridae (human crab louse) ....*Pthirus pubis*
- Body more than twice as long as wide; abdomen long, wider than thorax, with paratergal plates as sclerotized caps or lobes 7

7. Paratergal plates on segments IV–VI with apical margins free from abdomen (Figure 9.18); abdomen with a single transverse row of setae on each segment (1 genus, 16 spp. on cercopithecid Primates) ....Pedicinidae
- Some paratergal plates on segments III–VIII as sclerotized caps over apex of each lateral lobe (Figure 9.2), plate margins not free; abdominal setae not in distinct transverse rows (1 genus, 4 spp. on anthropoid Primates) ....Pediculidae 19

8. Head and thorax thickly covered with setae; abdomen with thickly set setae, spines, or scales; abdominal plates lacking (Figure 9.1) (5 genera, 12 spp. on Pinnipeds and aquatic Mustelidae) ....Echinophthiriidae
- Head and thorax with only a few setae; abdomen without scales and with or without plates 9

9. Abdominal cuticle scaly, with transverse rows of minute sclerotized points and few setae (Figure 9.5); abdominal spiracles present only on segment VIII (1 genus, 2 spp. on macroscelid Insectivora) ....Neolinognathidae
- Abdomen with numerous setae and with spiracles on more than one segment, usually on segments III–VII 10

10. Antenna 3-segmented; the basal segment large, subequal in length to the combined two distal segments and with ventral, posteriorly directed, stout hook; head and thorax strongly sclerotized (1 genus, 1 sp. on Dermoptera: flying lemurs and colugos)..Hamophthiriidae
- Antenna 4- or 5-segmented; the basal segment, less than half the length of the remaining distal segments; basal antennal segment usually unarmed ventrally 11

11. Forelegs smaller in size and with a weaker claw than other legs, which are subequal in size and similar in shape and armature; without detached ventral plates on abdominal segment II 12
- Forelegs and midlegs subequal in size and shape, both more slender and smaller than hindlegs; usually with a pair of small detached ventral plates on abdominal segment II (Figure 9.14)

(5 genera, 40 spp. on Sciuridae: squirrels, marmots, and prairie dogs) . . . . .Enderleinellidae

12. Forecoxae relatively close together at base; abdomen with distinct paratergal plates apically free from body margin (Figure 9.6) 13
   • Forecoxae widely separated from each other; abdomen without distinct paratergal plates, at most with small membranous tubercles posterior to each spiracle (3 genera, 22 spp. on Bovidae, Cervidae, Giraffidae, Equidae, Canidae, and Procaviidae) . . . . .Linognathidae 21

13. Thorax with a small, but distinct, notal pit; mesothoracic phragmata connected dorsally; abdomen membranous with paratergal plates on segments IV–VI (1 genus, 2 spp. on Equidae: donkeys and zebras) . . . . .Ratemiidae
   • Thorax without a distinct notal pit; mesothoracic phragmata weakly developed, not joined dorsally; abdomen with well-developed tergal and sternal plates and with paratergal plates not restricted to segments IV–VI 14

14. Second abdominal sternite with a lateral extension that articulates with paratergal plate (Figure 9.6) (5 genera, 132 spp. on Rodentia, Insectivora, Lagomorpha, and prosimian Primates – 117 spp. in the genus *Hoplopleura*) . . . . .Hoplopleuridae
   • Sternite of second abdominal segment not extended to articulate with paratergal plate (14 genera, 175 spp. on Rodentia, Lagomorpha, Insectivora, and prosimian Primates) . . . . .Polyplacidae 27

Haematopinus: **key to some representative species on livestock**[b]

15. Head at least twice as long as wide (Figure 9.19); thoracic sternite without anteromedian projection (Figures 9.21, 9.22) 16
   • Head less than twice as long as wide (length and width usually subequal); thoracic sternite with or without anteromedian projection 17

16. Thoracic sternite longer than wide, sternal pits off the sternite (Figure 9.21) (horse sucking louse) . . . .*H. asini*
   • Thoracic sternite wider than long, sternal pits on the sternite (Figure 9.22) (hog louse) . . . .*H. suis*

17. Thoracic sternite subquadrate in shape, without anteromedian projection (Figure 9.27) (water buffalo louse) . . . .*H. tuberculatus*
   • Thoracic sternite longer than wide, with anteromedian projection 18

18. Anteromedian projection blunt and rounded (Figure 9.23), male genital plate with 6 (5–7) anterior setae (Figure 9.7) (short-nosed cattle louse) . . . .*H. eurysternus*
   • Anteromedian projection elongate, acute (Figure 9.24); male genital plate with 4 anterior setae (cattle tail louse) . . . .*H. quadripertusus*

Pediculus: **key to species**

19. Spiracles of abdominal segments 3 to 5 each within a small, circular, sclerotized area; on chimpanzees . . . .*P. schaffi*
   • Abdominal spiracles 3 to 5 not in circular sclerite 20

20. All paratergal plates clearly lacking evidence of lateral lobes; normally occurring on man although at times to be found on monkeys of the family Cebidae and on gibbons in captivity (human body and head lice) .... *P. humanus, P. capitus*
- Paratergal plates of abdominal segments 5 to 6 bearing strong lateral lobes, both dorsally and ventrally (Figure 9.9) occurring on New World monkeys .... *P. mjobergi*

Key to some representative genera and species of Linognathidae

21. Abdominal spiracles borne in a slightly sclerotized tubercle that projects at least slightly from the body (Figure 9.13); usually without more than 8 setae per abdominal segment; on Cervidae and Bovidae (4 species) .... *Solenopotes*
- Abdominal spiracles not borne in sclerotized tubercles; 2 or more rows of abdominal setae per segment, on Bovidae, Giraffidae, and Canidae (24 to 30 species) .... *Linognathus* 22

22. Head nearly or quite as broad as long, usually shorter than antennae 23
- Head almost twice as long as broad or longer, usually appreciably longer than antennae 24

23. Thorax with 4 long dorsal setae; head slightly longer than broad; on dogs, foxes, and ferrets (dog sucking louse) .... *L. setosus*
- Thorax with only 2 long dorsal setae; head as broad as long; on legs and feet of sheep (sheep foot louse) .... *L. pedalis*

24. Head more than twice as long as broad; gonopods more or less truncate or sinuate posteriorly and bearing a hood or tooth near or at inner angle; abdominal setae long or short, but rather sparse always 25
- Head about twice as long as broad; gonopods rounded posteriorly 26

25. Abdominal setae long, many overlapping those of following rows; gonopods truncate, bearing a tooth near inner angle (Figure 9.25); apical lobes prominent, fringed with long setae; spiracles normal; on goats and sheep (goat sucking louse) .... *L. stenopsis*
- Most abdominal setae short, few overlapping those of following rows; gonopods set closer together with posterior margin emarginate, bearing a hook at inner angle (Figure 9.26); apical lobes prominent, set with a number of moderately long setae not arranged as a fringe; spiracles moderately large; on cattle (long-nosed cattle louse) .... *L. vituli*

26. Head acutely expanded laterally behind antennae (Figure 9.10) (goat sucking louse) .... *L. africanus*
- Head not acutely expanded behind antennae (Figure 9.11) (sheep sucking louse) .... *L. ovillus*

Key to some genera of Polyplacidae

27. Paratergal plates as small, sclerotized, thornlike projections on lateral hind margins of abdominal segments (Figure 9.20) (on lagomorphs, 6 spp.) .... *Haemodipsus*
- Paratergal plates large, at least half as long as thoracic sternite on more than one abdominal segment 28

28. Paratergal plate of second abdominal segment divided longitudinally into two plates, ventral paratergal plate with projecting point (Figure 9.12); on rodents in Heteromyidae (13 spp.) . . . .*Fahrenholzia*

● Paratergal plate of second abdominal segment not divided into two plates 29

29. Thoracic sternite emarginate posteriorly or subquadrate in shape (Figure 9.16); on Sciuridae and some Muridae (41 spp.) . . . .*Neohaematopinus*

● Thoracic sternite pointed posteriorly, or, if truncate, then first antennal segment much enlarged (Figure. 9.15) on many mice and rats (76 spp.) . . . .*Polyplax*

[a]Modified from Kim and Ludwig (1978).
[b]Modified from Meleney and Kim (1974).

Sucking louse–host association

Family of Anoplura	Genus	Number of spp. (as of 1978)	Mammalian host group
Echinophthiriidae	*Antarctophthirus*	6	Pinnipedia
	Echinophthirius	1	Phocidae
	Lepidophthirus	2	Phocidae
	Proechinophthirus	2	Otariidae
	Latagophthirus	1	Lutra (Mustelidae)
Enderleinellidae	*Enderleinellus*	43	Sciuridae
	Werneckia	3	Sciurinae
	Microphthirus	1	*Glaucomys*
	Atopophthirus	1	*Petaurista*
	Phthirunculus	1	*Petaurista*
Haematopinidae	*Haematopinus*	22	Suidae, Bovidae, Cervidae, Equidae
Hamophthiriidae	*Hamophthirius*	1	*Cynocephalus* (Cynocephalidae)
Hoplopleuridae	*Hoplopleura*	117	Rodentia, Lagomorpha
	Pterophthirus	5	Hystricomorpha (Rodentia)
	Haematopinoides	1	Talpidae (Insectivora)
	Ancistroplax	2	Soricidae (Insectivora)
	Schnizophthirus	7	Myomorpha (Rodentia)
Hybophthiridae	*Hybophthirus*	1	Tubulidentata
Linognathidae	*Linognathus*	3	Bovidae, Giraffidae, Canidae
	Solenopotes	10	Cervidae, Bovidae
	Prolinognathus	8	Procaviidae
Microthoraciidae	*Microthoracius*	4	Camelidae
Neolinognathidae	*Neolinognathus*	2	Macroscelididae (Insectivora)
Pecaroecidae	*Pecaroecus*	1	Tayasuidae
Pedicinidae	*Pedicinus*	16	Cercopithecidae (Primates)
Pediculidae	*Pediculus*	4	Cebidae, Pongidae, Hominidae
Polyplacidae	*Polyplax*	76	Rodentia, Insectivora
	Proenderleinellus	1	Muridae
	Fahrenholzia	13	Sciuromorpha (Rodentia)

Sucking louse–host association (*continued*)

Family of Anoplura	Genus	Number of spp. (as of 1978)	Mammalian host group
	Alenapthirus	1	Sciuridae
	Neohaematopinus	41	Sciuridae, other Rodentia, Insectivora
	Sathrax	1	Tupaiidae (Primates)
	Docophthirus	1	Tupaiidae (Primates)
	Lemurphthirus	3	Lorisidae (Primates)
	Lemurpediculus	2	Lemuridae (Primates)
	Phthirpediculus	2	Indriidae (Primates)
	Eulinognathus	23	Rodentia
	Ctenophthirus	1	Echimyidae (Hystricomorpha-Rodentia)
	Scipio	4	(Hystricomorpha-Rodentia)
	Haemodipsus	6	Leporidae (Lagomorpha)
Pthiridae	*Pthirus*	2	Hominidae, Pongidae
Ratemiidae	*Ratemia*	2	Equidae

Modified from Kim and Ludwig (1978).

References

Ferris, G. F. 1951. *The Sucking Lice.* Memoirs, Pacific Coast Entomological Society, vol. 1. San Francisco. 320 pp.

Kim, K. C. 1975. Specific antiquity of the sucking lice and evolution of otariid seals. *Rapports et Proces-verbaux des Réunions, Conseil international pour l'Exploration de la Mer 169*:544–9.

Kim, K. C., and H. W. Ludwig. 1978. The family classification of the Anoplura. *Syst. Entomol. 3*:249–84.

Meleney, W. P., and K. C. Kim. 1974. A comparative study of cattle-infesting *Haematopinus,* with redescription of *H. quadripertusus* Fahrenholz, 1916 (Anoplura: Haematopinidae). *J. Parasitol. 60*:507–22.

Stojanovich, C. J., and H. D. Pratt. 1965. *Key to Anoplura of North America.* U.S. Dept. of Health, Education and Welfare, Communicable Disease Center, Atlanta. 21 pp.

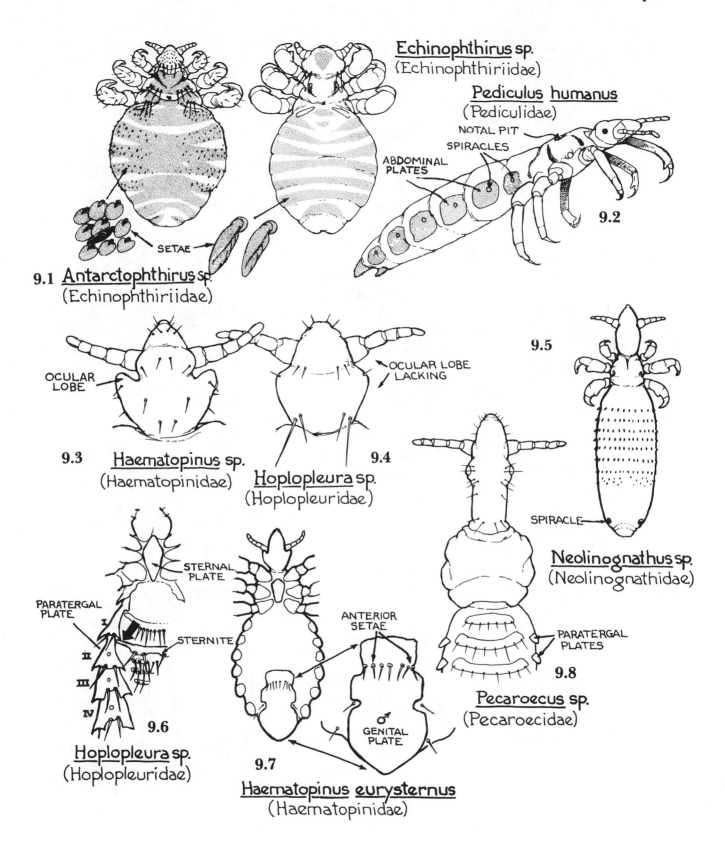

Echinophthirus sp.
(Echinophthiriidae)

Pediculus humanus
(Pediculidae)

NOTAL PIT

SPIRACLES

ABDOMINAL
PLATES

9.2

9.1 Antarctophthirus sp.
(Echinophthiriidae)

SETAE

9.5

OCULAR
LOBE

OCULAR LOBE
LACKING

9.3   Haematopinus sp.
(Haematopinidae)

9.4

Hoplopleura sp.
(Hoplopleuridae)

SPIRACLE

Neolinognathus sp.
(Neolinognathidae)

STERNAL
PLATE

PARATERGAL
PLATE

I

STERNITE

II

III

IV

9.6

Hoplopleura sp.
(Hoplopleuridae)

ANTERIOR
SETAE

♂

GENITAL
PLATE

PARATERGAL
PLATES

9.8

Pecaroecus sp.
(Pecaroecidae)

9.7

Haematopinus eurysternus
(Haematopinidae)

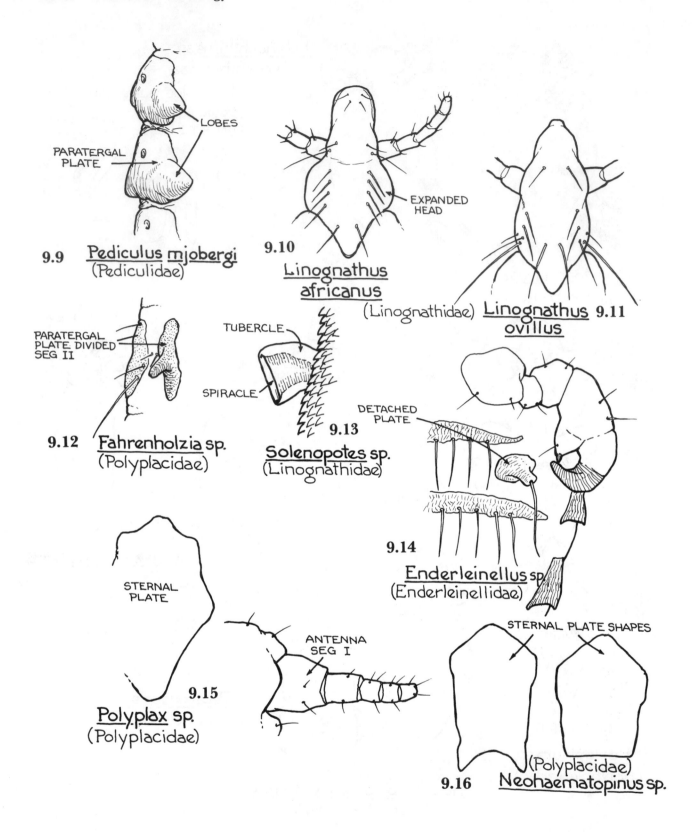

9.9 <u>Pediculus mjobergi</u>
(Pediculidae)

PARATERGAL PLATE

LOBES

9.10 <u>Linognathus</u>
<u>africanus</u>
(Linognathidae)

EXPANDED HEAD

<u>Linognathus</u> **9.11**
<u>ovillus</u>

PARATERGAL PLATE DIVIDED SEG II

9.12 <u>Fahrenholzia</u> sp.
(Polyplacidae)

TUBERCLE

SPIRACLE

9.13
<u>Solenopotes</u> sp.
(Linognathidae)

DETACHED PLATE

9.14
<u>Enderleinellus</u> sp.
(Enderleinellidae)

STERNAL PLATE

9.15
<u>Polyplax</u> sp.
(Polyplacidae)

ANTENNA SEG I

STERNAL PLATE SHAPES

(Polyplacidae)

9.16   <u>Neohaematopinus</u> sp.

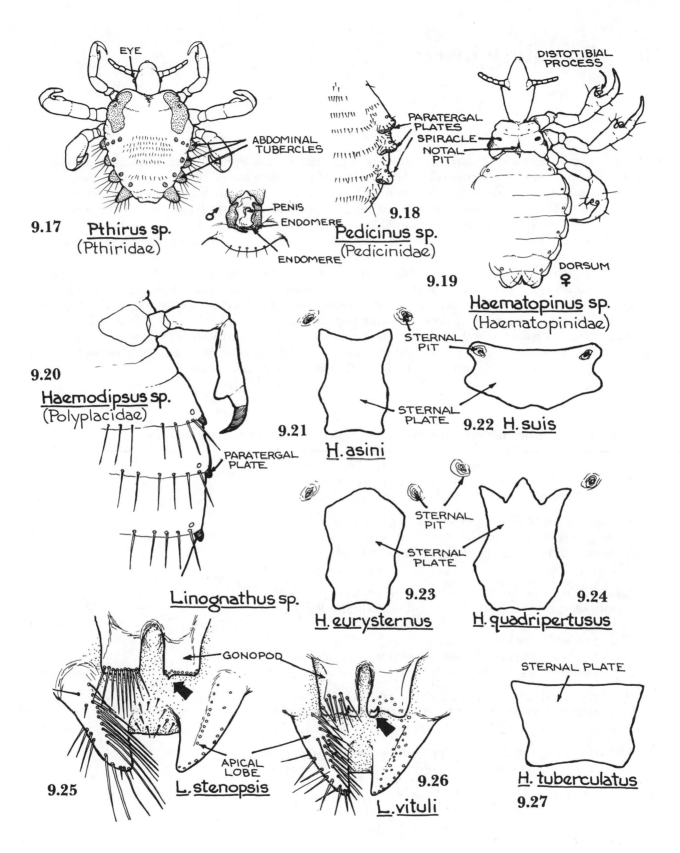

9.17   Pthirus sp. (Pthiridae)

EYE
ABDOMINAL TUBERCLES
♂ PENIS ENDOMERE ENDOMERE

9.18   Pedicinus sp. (Pedicinidae)

PARATERGAL PLATES
SPIRACLE
NOTAL PIT

9.19   Haematopinus sp. (Haematopinidae)

DISTOTIBIAL PROCESS
DORSUM ♀

9.20   Haemodipsus sp. (Polyplacidae)

PARATERGAL PLATE

Linognathus sp.

9.21   H. asini

STERNAL PLATE

9.22   H. suis

STERNAL PIT
STERNAL PLATE

9.23   H. eurysternus

STERNAL PIT
STERNAL PLATE

9.24   H. quadripertusus

9.25   L. stenopsis

GONOPOD
APICAL LOBE

9.26   L. vituli

STERNAL PLATE

H. tuberculatus
9.27

10   Immature insects

Among insects of medical importance, those with gradual metamorphosis, such as the lice or bugs, are characterized by a fairly close superficial resemblance of immature stages, or nymphs, and adults. Among holometabolous insects with complete metamorphosis, however, the young, or larvae, are very unlike the adults. Their different form is adapted to habitat and diet quite different from those of the adults. They occupy a far different ecological niche from that of the adults. In addition, the extreme change in form, or habitus, from larva to adult necessitates an intervening stage, the pupa, that prepares for the extreme metamorphic transformation.

With the exception of the maggots of higher flies as agents of myiasis, there are few immature stages of holometabolous insects parasitic on vertebrate hosts. However, the medical entomologist should become familiar with the larval characteristics of holometabolous insects of medical importance in order to determine the source and character of potential adult populations.

This exercise is divided into three parts: the first deals with the identification of immature holometabolous insects at the ordinal level, the second emphasizes identification of dipterous families, and the third deals with genera and species of Cyclorrhaphous flies of medical importance.

Key to orders of holometabolous insects of medical importance, late-stage larvae

1. Thorax with 3 pairs of legs (Figures 10.1, 10.3) 2
   * Thorax without true legs although 1 or more prolegs may be
   present (Figures 10.2, 10.7, 10.17) 4

2. Abdomen with at least 2 pairs of prolegs which are usually
   fleshy and seldom distinctly segmented (Figures 10.3, 10.5) 3
   * Abdomen without legs (do not confuse cerci with prolegs)
   (Figure 10.1) . . . . .Coleoptera

3. Abdomen with not more than 5 pairs of prolegs located on
   various segments but never on second segment; prolegs tipped
   with minute hooks (Figure 10.3, caterpillars) . . . . .Lepidoptera
   * Abdomen with more than 5 pairs of prolegs (Figure 10.5), 1
   pair of which occurs on the second segment; prolegs not
   tipped with minute hooks (sawfly "slugs") . . . . .Hymenoptera

4. Head capsule distinct, not peglike (Figures 10.4, 10.7, 10.8) 5
   * Lacking a distinct sclerotized head capsule, or, if apparently
   present, it is peglike (Figures 10.16, 10.17) 8

5. Abdomen with well-developed spiracles on several segments
   (or, if no spiracles evident on abdomen, head directed ventrad
   and the larva U-shaped) 6
   * Abdominal spiracles, if present, usually restricted to the
   terminal or subterminal segment; if present on other
   abdominal segments, spiracles rudimentary and
   inconspicuous; head directed cephalad 7

6. Abdominal segments with 1 or more lateral longitudinal folds
   or depressions extending from base to apex (weevil larvae)
   (Figure 10.2) . . . . .Coleoptera
   * Abdomen without such folds or depressions (Figure 10.6) . . . . .Hymenoptera

7. Abdomen with 9 or fewer segments, many with terminal gills
   or breathing tube (aquatic) (Figure 10.8); body form and size
   variable . . . . .Diptera
   ● Abdomen 10-segmented and with a pair of subanal processes;
   terrestrial and lacking gills or breathing tube; body cylindrical,
   segments bearing erect bristles (Figure 10.7) . . . . .Siphonaptera

8. Mandibles, if present, in the form of dark-colored hooks or
   sickle-shaped structures, moving vertically; body form
   variable, usually spindlelike, maggotlike, or grublike, body not
   normally curved in a U-shape (Figures 10.16, 10.17) . . . . .Diptera
   ● Mouthparts not as above; body U-shaped, stout, and largest
   near the middle (bee larvae) (Figure 10.6) . . . . .Hymenoptera

---

Key to suborders and families of Diptera of medical importance, last-instar larvae

1. Mandibles opposed (Figure 12.29), moving in a horizontal
   plane; head complete, or, if not, the posterior portion has deep
   longitudinal incisions . . . . .Nematocera 3
   ● Mandibles parallel, vertically biting; head poorly developed
   (Figures 10.16, 10.17) 2

2. Antennae well developed, though often small; maxillae and
   palpi well developed; mandibles usually sicklelike (Figures
   10.12, 10.14, 10.16) . . . . .Brachycera 12
   ● Antennae poorly developed or absent; maxillae poorly
   developed and palpi rarely visible; mandibles short and
   hooklike (Figure 10.23) . . . . .Cyclorrhapha 14

3. Head capsule complete, i.e., at least bounded behind and not
   divided into plates or rods 4
   ● Head incomplete behind, either with 3 wedge-shaped slits, 2
   on dorsum and 1 on ventral side, or the ventral surface very
   poorly sclerotized and the dorsal posteriorly in the form of 4
   slender, heavily sclerotized rods, with a weakly sclerotized
   plate on anterior half of dorsum (Figure 10.10) . . . . .Tipulidae

4. Prolegs (pseudopods) absent 5
   ● One or more prolegs present 8

5. The 3 thoracic segments separate; thorax and abdomen about
   equal in diameter 7
   ● The 3 thoracic segments fused to form a complex, more or less
   dilated mass thicker than abdomen (Figure 10.11) 6

6. Antennae prehensile (Figure 10.11), with long and strong
   apical spines . . . . .Chaoboridae
   ● Antennae not prehensile and lacking the strong apical spines
   (Figure 12.29) . . . . .Culicidae

7. Body without narrow dorsal plates or distinct secondary
   annuli; spiracles absent . . . . .Ceratopogonidae (= Heleidae) (in part)
   ● Dorsal surface of segments with narrow transverse sclerotized
   bands; thorax and abdomen with secondary segmentation;
   spiracles on prothorax and anal segment . . . . .Psychodidae (in part)
   Psychodinae

8. Prolegs confined to anterior or posterior ends of body, or both 10
 ● Not as above 9

9. Paired prolegs situated on abdominal segments 1 and 2 only,
   and each provided with many small hooks; posterior end of
   body with 2 ciliated lateral lobes and a terminal bristly lobe ....Dixidae
 ● Most abdominal segments with a single padlike proleg,
   sometimes slightly bilobed; posterior end of body not as above
   (Figure 10.13) ....Psychodidae (in part)
   Phlebotominae

10. Posterior end of body swollen and club-shaped; proleg
    restricted to prothorax; body with caudal ring of hooks for
    clinging to objects in running water; maxillae with a fan of
    long hairs (Figure 10.8) ....Simuliidae
 ● Posterior end of body not swollen and club-shaped; prolegs
   present on both prothorax and posterior end of body or at
   posterior end only; lacking a caudal ring of hooks; maxillae
   lacking long fanlike hairs 11

11. Prolegs normally present anteriorly and posteriorly; if
    posterior pair reduced, the anterior pair at least represented
    by patches of spinules; body hairs weakly developed (Figure
    10.9) ....Chironomidae (= Tendipedidae)
 ● If anterior and posterior prolegs present, the body has strong
   spines or bristles; if anterior prolegs missing, the posterior
   proleg is represented by a few ventral claws, and the body lacks
   spines or bristles ....Ceratopogonidae (= Heleidae) (in part)

12. Body entirely shagreened or longitudinally striated, at least in
    part; posterior spiracles usually concealed in a cleft,
    approximated (Figure 10.16) 13
 ● Body not shagreened or visibly striated (Figure 10.14),
   posterior spiracles visible or concealed in a horizontal cleft,
   rather widely separated (Figure 10.15) ....Athericidae, Rhagionidae

13. Head not retractile (Figure 10.12), body flattened, shagreened,
    bristly, without prolegs (pseudopods) ....Stratiomyidae
 ● Head retractile (Figure 10.16), body cylindrical; not
   shagreened but finely longitudinally striated; abdomen with a
   girdle of prolegs on each segment (Figure 10.16) ....Tabanidae

Genera and species of Cyclorrhapha of medical importance in the Americas, last-instar larvae[a]

14. Posterior spiracles on fleshy or sclerotized process or on an
    elongate tube 15
 ● Posterior spiracles flush with cuticular surface or in a sunken
   cavity (not on an elongated or raised process) 21

15. Flattened larvae with fleshy or spinous dorsal and lateral
    processes; posterior spiracles on separate fleshy stalks or
    sclerotized tubercles (Muscidae, Faniinae; Phoridae) 16
 ● Cylindrical or grublike larvae; body smooth or with short
   dorsal and lateral tubercles or spines; posterior spiracles may
   be on a common stalk 18

16. Posterior spiracles borne on a pair of sclerotized tubercles;
    dorsal and lateral processes short and unbranched ....Phoridae

- Posterior spiracles borne on a pair of apically branched, fleshy stalks (Figure 10.18); dorsal and lateral processes branched basally . . . . .Muscidae (in part), *Fannia* 17

17. Lateral and dorsal processes with short, broad spines on basal portion (Figure 10.19) (lesser house flies) . . . .*Fannia canicularis*
- Lateral and dorsal processes "feathery," branched (Figure 10.20) (latrine flies) . . . .*Fannia scalaris*

18. Posterior spiracles on paired, peglike processes or cones that are much shorter than the body width; anal segment with a number of short fleshy processes 19
- Posterior spiracles on a common, unforked, tubular, taillike process that is much longer than body width (may be longer than the body length when fully extended); some aquatic . . . . .Syrphidae

19. Posterior spiracles on a pair of short, strongly tapered cones that point upward (Figure 10.22); anterior spiracle fanlike. Mature larvae move in a skipping manner (cheese skippers) . . . .*Piophila* Piophilidae
- Posterior spiracles on a pair of basally contiguous, short, peglike processes directed posteriorly (Figure 10.21); anterior spiracles on a lateral process or not fanlike 20

20. Anterior spiracle on short, lateral, peglike process; body without cuticular hairs or spines (vinegar or fruit flies) . . . . .Drosophilidae
- Anterior spiracle not on a lateral process, with lobes arising from a central column; body with cuticular hairs or spines . . . . .Sepsidae

21. Peritreme (sclerotized ringlike structure surrounding each posterior spiracle) present; posterior spiracle with 3 distinct slits (Figures 10.17, 10.27) 22
- Peritreme absent (Figure 10.35); or if present, spiracle without 3 distinct slits 31

22. Slits of posterior spiracles straight (Figure 10.27) 23
- Slits of posterior spiracles strongly sinuous (Figure 10.34) .Muscidae (in part) 30

23. Dorsal and ventral arms of cephaloskeleton extend a subequal distance (Figure 10.25); peritreme very thin on two areas away from the button (Figure 10.26) (black garbage flies) . . . .(*Ophyra*) Muscidae (in part)
- Dorsal arm of cephaloskeleton longer than ventral arm (Figure 10.28); peritreme complete or with one weakly sclerotized or incomplete area (Figures 10.24, 10.38) 24

24. Posterior spiracles with peritreme complete, sometimes weak in area of button (Figures 10.27, 10.31) 25
- Posterior spiracles with peritreme incomplete, not enclosing a sometimes ill-defined button (Figures 10.24, 10.30) 26

25. Accessory oral sclerite present (Figure 10.28); posterior spiracular plate and button heavily sclerotized (blue bottle flies) . . . . *Calliphora* and *Cynomyopsis* of Calliphoridae
- Accessory oral sclerite absent (Figure 10.29) spiracular plate and button not heavily sclerotized (Figure 10.31) (green bottle flies) . . . .*Phaenicia* spp. of Calliphoridae

26. Inner (medial) slits of posterior spiracles not directed toward ventral midline (Figure 10.30) . . . . .Sarcophagidae (in part) 27

- Inner slits of posterior spiracles directed toward ventral midline (Figure 10.24) . . . . .Calliphoridae 28

27. Inner slits not pointing toward opening in peritreme (Figure 10.30) . . . .*Sarcophaga*
- Inner slits pointing toward opening in peritreme . . . .*Wohlfahrtia*

28. Spiracular button present; walls of slits without lateral swellings (Figure 10.24) (black blow flies) . . . .*Phormia regina*
- Spiracular button indistinct or absent; walls of slits with lateral swellings 29

29. Tracheal trunks leading from posterior spiracles pigmented (Figure 10.32) (primary screw worms) . . . .*Cochliomyia hominivorax*
- Tracheal trunks leading from posterior spiracles not pigmented (Figure 10.33) (secondary screw worms) . . . .*Cochliomyia macellaria*

30. Peritreme thick (Figure 10.17). Found in wide range of decaying organic debris (house flies) . . . .*Musca domestica*
- Peritreme thin (Figure 10.34). Primarily in fresh cattle droppings (horn flies) . . . .*Haematobia irritans*

31. Cylindrical, slender larvae, usually less than 13 mm long, tapering anteriorly (as in Figure 10.17) . . . . .Muscidae (in part) 32
- Robust, grublike larvae over 15 mm long, often with very stout spines (Figures 10.43, 10.44) 34

32. Posterior spiracular button centrally placed (Figure 10.35) (stable flies) . . . .*Stomoxys calcitrans*
- Posterior spiracular button not centrally placed (Figure 10.36) 33

33. Slits of posterior spiracles strongly sinuous (Figure 10.37) (face flies) . . . .*Musca autumnalis*
- Slits of posterior spiracles not strongly sinuous (Figure 10.36) (false stable flies) . . . .*Muscina stabulans*

34. Three distinct slits in each posterior spiracle (Figure 10.44) 35
- Posterior spiracles lacking 3 distinct slits but with numerous small openings, some of which may be sinuous (Figure 10.40) 36

35. Slits of posterior spiracles very long, bent at the middle; body ovate (horse bot stomach flies) (Figure 10.44) . . . . *Gasterophilus* Gasterophilidae
- Slits of posterior spiracles long and relatively straight; body enlarged anteriorly, club-shaped when extended (Figure 10.39) (torsalo bot fly of livestock and man, Central and S. America) . . . .*Dermatobia hominis* Cuterebridae (in part)

36. Mouth hooks vestigial (cattle grubs) . . . . .Oestridae, Hypodermatinae 37
- Mouth hooks well developed 38

37. Posterior spiracle with opening toward button narrow; stigmatal plate sloping centrally, funnellike toward button (Figure 10.42) (northern cattle grub) . . . .*Hypoderma bovis*
- Posterior spiracle with opening toward button wide, stigmatal plate not sloping funnellike toward the button (Figure 10.41) (common cattle grub) . . . .*Hypoderma lineatum*

38. Body spines weak, restricted to the venter and to anterior margins of segments dorsally (sheep nasal bot fly) . . . .*Oestrus ovis* Oestridae (in part)

● Body with strong stout spines generally distributed 39

39. Spiracular slits very sinuous (Figure 10.43); body densely
covered with stout spines (rodent and rabbit bot flies) . . . . .Cuterebridae (in part) 40
   ● Spiracular plates densely perforated (Figure 10.40); body
spines not completely covering integument (deer nasal bot
flies) . . . .*Cephenemyia* Oestridae (in part)

40. Body spines multipointed (rodent bot flies) . . . .*Cuterebra* (in part)
   ● Body spines single, thornlike, not multipointed (rabbit bot
flies) . . . .*Cuterebra* (in part)

^aAdapted from Stojanovich et al. (1966) and Oldroyd and Smith (1973).

References

Chu, H. F. 1949. *How to Know the Immature Insects.* Wm. C. Brown, Dubuque, Iowa. 234 pp.

Hennig, W. 1948–52. *Die Larvenformen der Dipteren,* Teil I–III. Akademie Verlag, Berlin.

James, M. T. 1947. *The Flies That Cause Myiasis in Man.* U.S. Dept. Agr. Misc. Publ. 631. 175 pp.

Merritt, R. W., and K. W. Cummins. 1978. *An Introduction to Aquatic Insects of North America.* Kendall/Hunt Publ., Dubuque, Iowa. 441 pp.

Oldroyd, H., and G. U. Smith. 1973. Eggs and larvae of flies. In *Insects and Other Arthropods of Medical Importance,* ed. K. G. V. Smith. British Museum (Natural History), London, pp. 289–323.

Peterson, A. 1951. *Larvae of Insects,* vols. I and II. Edwards Bros., Ann Arbor, Mich. 731 pp.

Stojanovich, C. J., H. D. Pratt, and E. E. Bennington. 1966. Fly larvae: key to some species of public health importance. In *Pictorial Keys to Arthropods, Reptiles, Birds and Mammals of Public Health Significance.* U.S. Department of Health, Education, and Welfare, Communicable Disease Center, Atlanta, pp. 125–33.

Teskey, H. J. 1969. Larvae and pupae of some eastern North American Tabanidae (Diptera). *Mem. Entomol. Soc. Canada* No. *63*:1–147.

Usinger, R. L. 1956. *Aquatic Insects of California.* University of California Press, Berkeley. 508 pp.

Zumpt, F. 1965. *Myiasis of Man and Animals in the Old World.* Butterworth's, London. 267 pp.

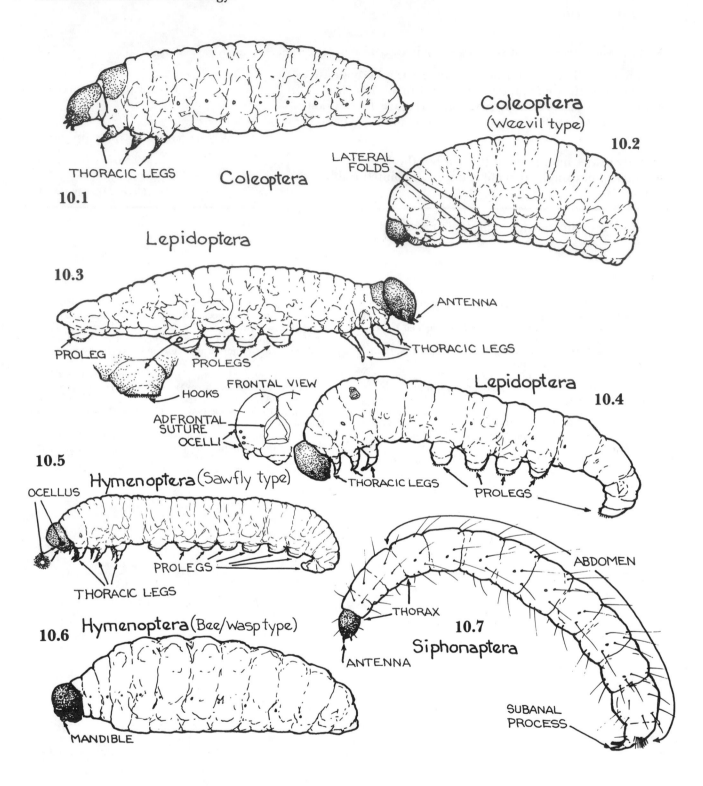

THORACIC LEGS

Coleoptera

10.1

Coleoptera
(Weevil type)

10.2

LATERAL
FOLDS

Lepidoptera

10.3

PROLEG

PROLEGS

HOOKS

FRONTAL VIEW

Lepidoptera

10.4

ADFRONTAL
SUTURE

OCELLI

THORACIC LEGS

PROLEGS

10.5

OCELLUS

Hymenoptera (Sawfly type)

PROLEGS

THORACIC LEGS

ABDOMEN

THORAX

10.7
Siphonaptera

ANTENNA

10.6

Hymenoptera (Bee/wasp type)

MANDIBLE

SUBANAL
PROCESS

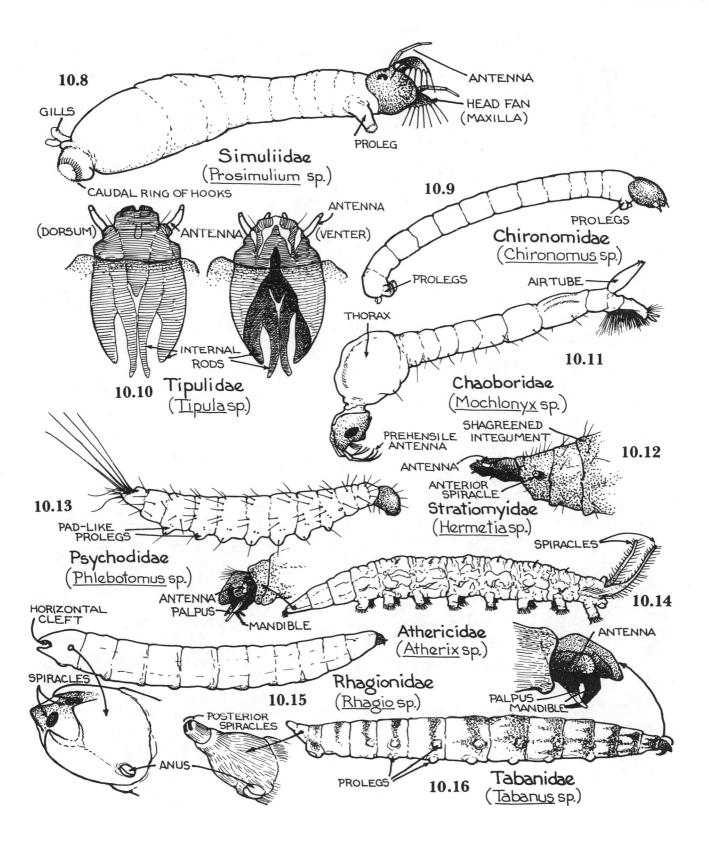

10.8
ANTENNA
HEAD FAN
(MAXILLA)
GILLS
PROLEG
Simuliidae
(_Prosimulium_ sp.)
CAUDAL RING OF HOOKS

(DORSUM)
ANTENNA
ANTENNA
(VENTER)
INTERNAL
RODS
10.10 Tipulidae
(_Tipula_ sp.)

10.9
PROLEGS
Chironomidae
(_Chironomus_ sp.)
PROLEGS
AIR TUBE
THORAX
10.11
Chaoboridae
(_Mochlonyx_ sp.)
PREHENSILE
ANTENNA
SHAGREENED
INTEGUMENT
ANTENNA
ANTERIOR
SPIRACLE
10.12
Stratiomyidae
(_Hermetia_ sp.)

10.13
PAD-LIKE
PROLEGS
Psychodidae
(_Phlebotomus_ sp.)
ANTENNA
PALPUS
MANDIBLE
SPIRACLES
10.14
Athericidae
(_Atherix_ sp.)
ANTENNA
PALPUS
MANDIBLE

HORIZONTAL
CLEFT
SPIRACLES
ANUS
Rhagionidae
(_Rhagio_ sp.)
10.15
POSTERIOR
SPIRACLES
PROLEGS
10.16
Tabanidae
(_Tabanus_ sp.)

SPIRACULAR SLITS

THORAX

ANTERIOR SPIRACLE

BUTTON

MID-LINE

PERITREME

10.17

MOUTH HOOK

ANTENNA

Musca sp.
(Muscidae)

SPIRACULAR STALK

DORSAL PROCESSES

10.19
F. canicularis
(Muscidae)

THORAX

10.18

LATERAL PROCESSES

Fannia sp.
(Muscidae)

10.20
F. scalaris
(Muscidae)

PEG-LIKE PROCESS

CONE PROCESS

Piophila sp.
(Piophilidae)

Calliphoridae

ANTERIOR SPIRACLE

10.21

10.22

10.23

Drosophila sp.
(Drosophilidae)

INNER SPIRACULAR SLIT

DORSAL ARM

Ophyra sp.

10.24   Phormia sp.
(Calliphoridae)

BUTTON

MID-LINE

MID-LINE

VENTRAL ARM

Ophyra sp.

10.25

10.26

ACCESSORY SCLERITE

10.29

10.27

Calliphora sp.

10.28

Phaenicia sp.

MID-LINE

INNER SPIRACULAR SLIT

10.30   Sarcophagidae

ANTERIOR SPIRACLE

MID-LINE

10.31

MID-LINE

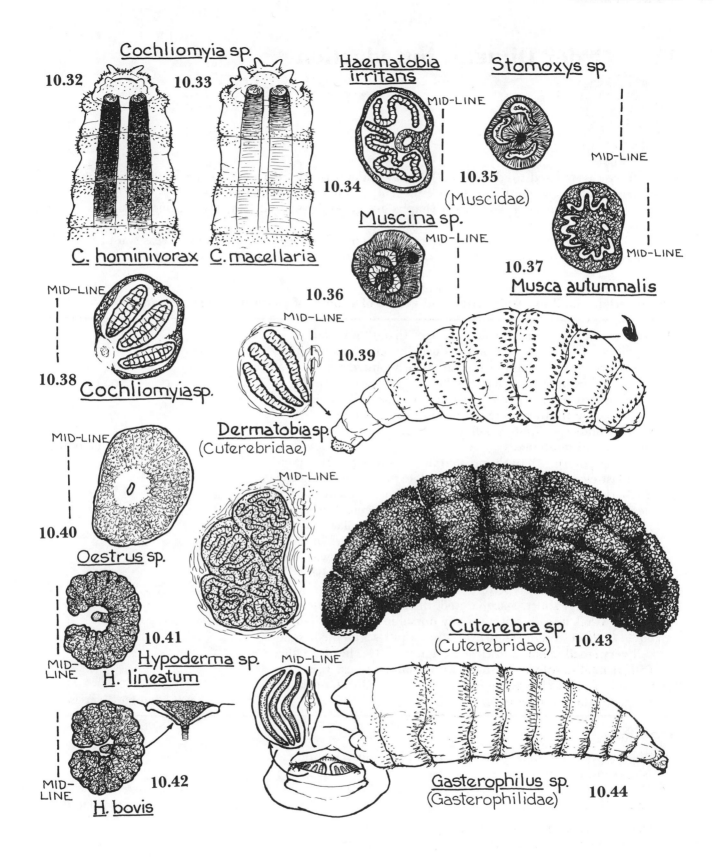

Cochliomyia sp.

10.32   10.33

C. hominivorax   C. macellaria

Haematobia irritans

MID-LINE

10.34

Stomoxys sp.

MID-LINE

MID-LINE

10.35
(Muscidae)

Muscina sp.

MID-LINE

MID-LINE

10.36

10.37

Musca autumnalis

MID-LINE

10.38   Cochliomyia sp.

MID-LINE

10.39

Dermatobia sp.
(Cuterebridae)

MID-LINE

10.40

Oestrus sp.

MID-LINE

MID-LINE

10.41   Hypoderma sp.
H. lineatum

MID-LINE
MID-LINE

10.42
H. bovis

MID-LINE

Cuterebra sp.
(Cuterebridae)   10.43

MID-LINE

Gasterophilus sp.
(Gasterophilidae)   10.44

11   Order Diptera: introduction

This exercise is devoted to identifying Nematocera and Brachycera with the aid of the key given below. Other classifications within the order are presented in following exercises.

Nematocera, Brachycera, and Cyclorrhapha are generally winged, but some forms in each group are wingless or possess rudimentary wings as adults.

The wing membranes of Diptera are strengthened by veins (see figures of wings). Areas bounded by cross veins, long veins, or the wing margin are called cells. A cell entirely bounded by veins is said to be "closed."

Key to certain families, subfamilies, and genera of Diptera of medical importance, adults

1. Antennae many-jointed, usually longer than thorax, but segments from number 3 onward usually very similar and frequently closely joined so as to simulate 3-jointed antennae of the following group (Figures 11.1, 11.11); arista wanting; palpi usually 4- or 5-jointed, pendulous; anal cell (cubital) widely open or absent (Figure 12.7) (these flies mostly slender-bodied or quite small, of the type of crane flies, midges, and mosquitoes) ....Nematocera 2
   - Antennae shorter than thorax, variable, usually 3-segmented with last elongate (last segment annulate in tabanids) (Figure 16.3), thus superficially resembling the many-segmented type described above; arista terminal (rarely subterminal) when present; palpi 1- or 2-segmented, porrect; anal cell contracted before wing margin or closed (Figure 16.2) (the flies in this group mostly medium to large, usually stout-bodied, with the posterior segments often produced into a point or tube; large-winged, of the type of deer flies and horseflies) ....Brachycera 28
   - Antennae very short, 3-segmented, the last segment often greatly swollen, with arista usually dorsal in position (Figures 17.1, 17.2) or absent (Figure 17.26); palpi 1-segmented; anal cell very small, basal, and closed (Figure 17.15); head usually with frontal lunule and usually with ptilinum (these flies variable, small to large, but usually stout-bodied with medium wings, of the muscoid type) ....Cyclorrhapha (Exercise 17)

Nematocera

2. Mesonotum without a V-shaped suture; if a transverse suture is present, it is incomplete 3
   - Mesonotum with an entire V-shaped suture (Figure 11.1) (crane flies) ....Tipulidae

3. Ocelli present; antennae usually shorter than thorax 4
   - Ocelli absent; antennae usually longer than thorax 5

4. Antennae arising ventrad of the eyes (March flies) ....Bibionidae
   - Antennae arising at level of middle of eyes or above (fungus gnats) ....Mycetophilidae

5. Costal vein ending at or near apex of wing (Figure 11.5) 6
 • Costal vein continuing around margin beyond apex of wing
   although often weaker along rear margin of wing 8

6. Medius 3 and 4 and cubitus fused to about midwing where
   they fork (Figure 11.9); wings relatively narrow basally 7
 • Anterior veins thick, others very weak; medius 3 and 4 and
   cubitus separate to their base; wings broad and short (Figure
   11.5) (black flies) . . . . .Simuliidae 11

7. Mouthparts not fitted for piercing; postscutellum (postnotum
   of authors) usually with median longitudinal groove (Figure
   11.2) (midges) . . . . .Chironomidae (= Tendipedidae)
 • Mouthparts fitted for piercing; postscutellum rounded,
   without median groove (no-see-ums) . . . . .Ceratopogonidae (= Heleidae) 22

8. Wing veins very hairy or scaly (Figures 11.12, 12.7); body and
   legs hairy or scaly 9
 • Wing veins and body lacking scales . . . . .Dixidae

9. Wings hairy, short, broadly ovate or pointed; cross veins absent
   except near base of wing (Figure 11.14) (moth flies) . . . . .Psychodidae 16
 • Wings scaly, longer, and narrow; cross veins not limited to base
   of wing 10

10. Proboscis short, about one-third length of thorax;
    sternopleuron with transverse suture (Figure 11.3, 11.17)
    (phantom midges) . . . . .Chaoboridae 25
 • Proboscis long, as long as or longer than thorax (Figure 12.4);
   sternopleuron (mesepisternum) without transverse suture
   (Figure 12.1) (mosquitoes) . . . . .Culicidae (Exercise 12)

Genera of North American Simuliidae, adults[a]

11. Scutum with stout, erect hairs but no fine recumbent hairs;
    antenna 9-segmented; a bulla (blisterlike structure) behind eye
    laterally . . . .*Gymnopais*
 • Scutum usually with fine recumbent hairs but never stout,
   erect hairs; antenna with 9–11 segments (Figure 11.5); with or
   without a bulla behind eye laterally 12

12. Costa with fine hairs only, not interspersed with spinules;
    radial sector distinctly forked apically (Figure 11.7); with or
    without a bulla behind eye laterally 13
 • Costa usually with spinules interspersed among the fine hairs;
   radial sector simple (occasionally obscurely forked at extreme
   apical portion); no bulla behind eye 15

13. Vein R_1 joining costa at about middle of wing; fork of radial
    sector ending before termination of costa; m-cu fold
    apparently unforked; vein Cu_2 nearly straight . . . .*Parasimulium*
 • Vein R_1 joining costa well beyond middle of wing; fork of
   radial sector ending near termination of costa; m-cu fold
   forked apically; vein Cu_2 sinuous (Figure 11.7) 14

14. Antenna 9-segmented; at least an indication of a bulla behind
    eye (Figure 11.6) . . . .*Twinnia*
 • Antenna 10- or 11-segmented (9-segmented in *P. gibsoni*); no
   bulla behind eye . . . .*Prosimulium*

15. Length of vein R not less than one-third the remaining distance to apex of wing, with hair dorsally; basal cell of wing usually distinguishable (Figure 11.8); second hind tarsal segment without pedisulcus, or this represented by a shallow depression only (includes *Greniera, Stegopterna, Metacnephia, Ectemnia,* and *Cnephia*) . . . . Cnephiini
   - Length of vein R equal to much less than one-third the remaining distance to apex of wing, with or without hair dorsally; basal cell of wing absent; second hind tarsal segment with a distinct, usually deep pedisulcus (Figure 11.5) . . . . *Simulium*

Selected genera of New World Psychodidae[b]

16. Eyes circular in outline; flagellar segments of antenna pyriform or subcylindrical 17
   - Eyes with a median extension (Figure 11.13); flagellar segments of antenna barrel-shaped or nodelike . . . .(Psychodinae) 21

17. One longitudinal vein between R 2+3 and M 1+2 (i.e., R-sector is 3-branched) (Figure 11.15); labellum bulbous . . . .(Trichomyiinae) *Trichomyia*
   - Two longitudinal veins between R 2+3 and M 1+2 (i.e., R-sector is 4-branched) (Figure 11.14); labellum elongated, adapted for piercing . . . .(Phlebotominae = Phlebotomidae of some authors) 18

18. Third maxillary palpal segment longer than the 5th (Figure 11.12); abdominal setae recumbent on V and VI (dominant biter of humans in neotropical, highland forests) . . . . *Psychodopygus* (28 spp.)
   - Third maxillary palpal segment shorter than the 5th, the longest; abdominal segments V and VI with erect setae 19

19. Mesanepisternal setae absent; wing shape broad and rounded at tip . . . . *Warileya* (5 spp.)
   - Mesanepisternal setae present (in two groups); wing shape relatively narrow and pointed at tip (Figure 11.14) 20

20. Female cibarial armature with horizontal teeth only, typically in 4 longitudinal rows . . . .*Brumptomyia* (22 spp.)
   - Female cibarial armature with 1 row of horizontal teeth and often with comb and vertical teeth . . . .*Lutzomyia* (263 spp.)

21. Terminal flagellar segments reduced in size (only about one-half the size of proximal segments) (Figure 11.13) . . . .*Psychoda*
   - Terminal flagella segments subequal or larger in size to proximal segments . . . .*Telmatoscopus*

Medically important genera of North American Ceratopogonidae, adults[c]

22. Media vein not forked, r-m cross vein absent, wings without macrotrichia . . . .*Leptoconops*
   - Media vein forked, r-m cross vein present, wings with some macrotrichia (Figure 11.9) 23

23. Empodium large, well developed, wings with numerous macrotrichia . . . .*Forcipomyia*
   - Empodium small or vestigial 24

24. First radial cell small or absent, second radial cell absent or, if present, square-ended and ending before middle of wing (Figure 11.10) . . . .*Dasyhelea*

- Both radial cells present, second cell not square-ended and ending beyond middle of wing (Figure 11.9), wing usually patterned . . . . *Culicoides*

Genera of North American Chaoboridae, adults

25. First tarsal segment longer than second 26
- First tarsal segment shorter than second (Figure 11.18) . . . .*Mochlonyx*

26. Anal vein terminates at or before fork of vein Cu . . . .*Eucorethra*
- Anal vein terminates distad of fork of vein Cu 27

27. Vein R_1 terminates closer to R_2 than to Sc (Figure 11.17) . . . .*Chaoborus*
- Vein R_1 terminates closer to Sc than to R_2 (Figure 11.16) . . . .*Corethrella*

Brachycera

28. Tarsi with 3 nearly equal pads under tarsal claws (empodium developed pulvilliform); vein R_{4-5} forked (Figure 16.2) 29
- Tarsi with 2 pads under claws (empodium hairlike or absent); vein R_{4-5} not forked (long-legged flies) . . . . .Dolichopodidae

29. Calypters large, conspicuous; third antennal segment annulated into 3–8 apparent segments, rarely bearing an elongate arista (Figure 16.1) 30
- Calypters small, vestigial; third antennal segment compact, not annulated, usually bearing an elongate arista (Figure 11.4) (snipe flies) . . . . .Athericidae, Rhagionidae

30. Apical tibial spurs present at least on midtibia; costa continuing around hind margin of wing; squamae large and conspicuous (Figure 16.2) (horse flies and deer flies) . . . . .Tabanidae (Exercise 16)
- Apical tibial spurs absent; costa ending before wing tip; squamae small or vestigial (soldier flies) . . . . .Stratiomyidae

[a]From Peterson (1960).
[b]Modified from Quate (1955).
[c]Modified from Freeman (1973).

References

California Mosquito Control Association, Entomology Committee. 1967. *A Field Guide to Common Mosquito-like Gnats of California.* California Mosquito Control Association, Visalia.

Colyer, C. N., and C. O. Hammond. 1951. *Flies of the British Isles.* Warne, New York.

Curran, C. H. 1934. *The Families and Genera of North American Diptera.* American Museum of Natural History, New York.

Davies, D. M., B. V. Peterson, and D. M. Wood. 1962. The black flies (Diptera: Simuliidae) of Ontario. I. Adult identification and distribution with descriptions of six new species. *Proc. Entomol. Soc. Ontario* 92:70–154.

Edwards, F. W., H. Oldroyd, and J. Smart. 1939. *British Blood-Sucking Flies.* British Museum of Natural History, London.

Freeman, P. 1973. Ceratopogonidae (biting midges, sandflies, punkles). In *Insects and Other Arthropods of Medical Significance,* ed. K. G. V. Smith. British Museum of Natural History, London. 561 pp.

Peterson, B. V. 1960. The Simuliidae (Diptera) of Utah. I. Keys, original citations, types and distribution. *Great Basin Naturalist* 20:81–104.

Quate, L. W. 1955. A revision of the Psychodidae (Diptera) in America North of Mexico. *U. of Cal. Pub. Entomol.* 10(3):103–273.

Ready, P. D., H. Fraiha, R. Lainson, and J. J. Shaw. 1980. *Psychodopygus* as a genus: reasons for a flexible classification of the phlebotomine sand flies. *J. Med. Entomol.* 17:75–88.

Stone, A., C. W. Sabrosky, W. W. Wirth, R. H. Foote, and J. R. Coulson. 1965. *A Catalog of the Diptera of America North of Mexico.* Agriculture Handbook No. 276. U.S. Gov. Print. Office, Washington, D.C.

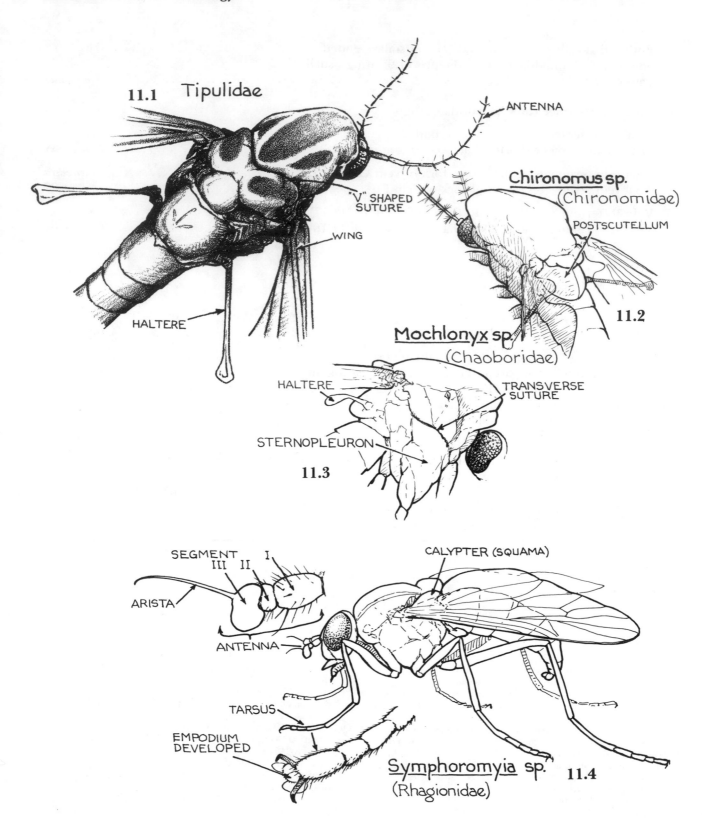

11.1 Tipulidae

ANTENNA

"V" SHAPED SUTURE

WING

HALTERE

Chironomus sp. (Chironomidae)

POSTSCUTELLUM

11.2

Mochlonyx sp. (Chaoboridae)

HALTERE

TRANSVERSE SUTURE

STERNOPLEURON

11.3

SEGMENT III II I

CALYPTER (SQUAMA)

ARISTA

ANTENNA

TARSUS

EMPODIUM DEVELOPED

Symphoromyia sp. (Rhagionidae) **11.4**

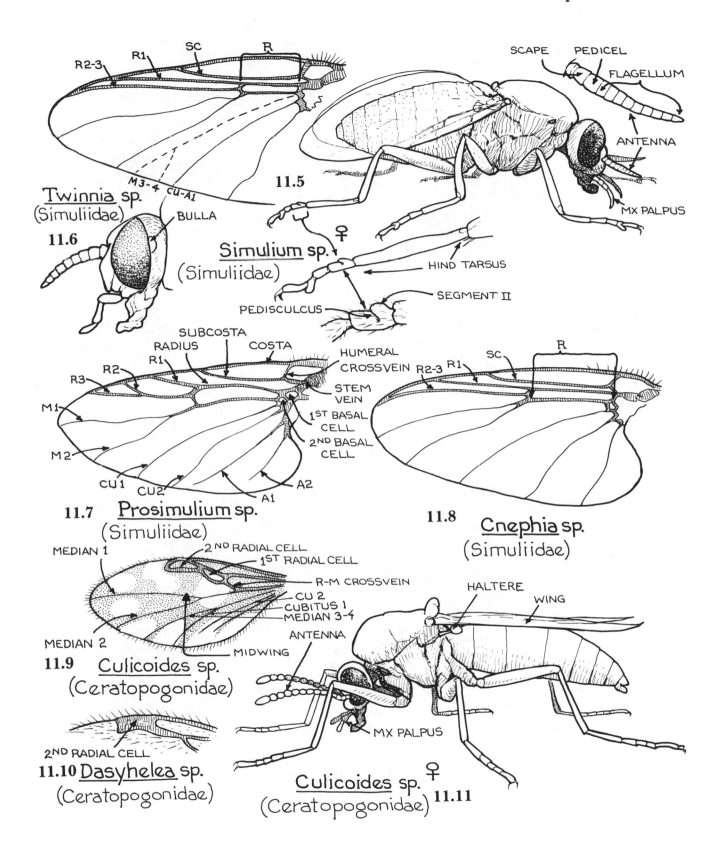

11.5 Simulium sp. ♀ (Simuliidae)

11.6 Twinnia sp. (Simuliidae) — BULLA

11.7 Prosimulium sp. (Simuliidae)

11.8 Cnephia sp. (Simuliidae)

11.9 Culicoides sp. (Ceratopogonidae)

11.10 Dasyhelea sp. (Ceratopogonidae)

11.11 Culicoides sp. ♀ (Ceratopogonidae)

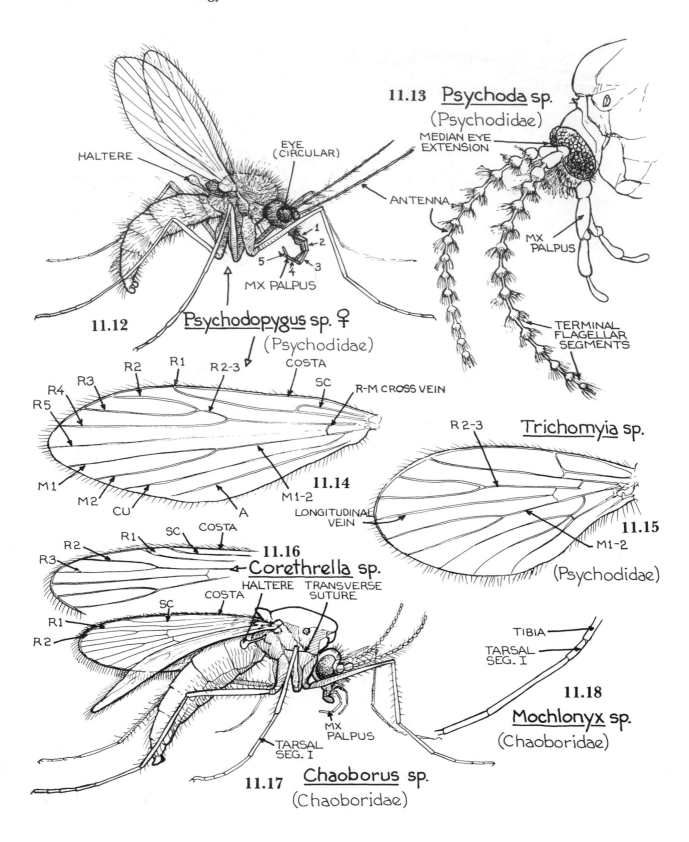

11.13 <u>Psychoda</u> sp.
(Psychodidae)

MEDIAN EYE
EXTENSION

ANTENNA

MX
PALPUS

TERMINAL
FLAGELLAR
SEGMENTS

HALTERE

EYE
(CIRCULAR)

1
2
5
3
4
MX PALPUS

11.12 <u>Psychodopygus</u> sp. ♀
(Psychodidae)

R2   R1   R2-3   COSTA
R3   SC
R4   R-M CROSS VEIN
R5

M1
M2   CU   A   M1-2
          11.14

R 2-3   <u>Trichomyia</u> sp.

LONGITUDINAL
VEIN
                              11.15
                    M1-2
(Psychodidae)

R1   SC   COSTA
R2   **11.16**
R3   <u>Corethrella</u> sp.

COSTA   HALTERE   TRANSVERSE
SC   SUTURE
R1
R2
                    MX
                    PALPUS
TARSAL
SEG. I
          11.17 <u>Chaoborus</u> sp.
          (Chaoboridae)

TIBIA
TARSAL
SEG. I
          11.18
          <u>Mochlonyx</u> sp.
          (Chaoboridae)

12 Order Diptera: Culicidae to genus (adults, pupae, larvae); Chaoboridae to genus (larvae)

Mosquitoes, the Culicidae, are arthropods of preeminent importance to medical entomology. Both as vectors of disease agents and through their direct attacks they make some localities seasonally uninhabitable. In no other group of animals has the taxonomy for all stages been so completely studied. There are more than 2,500 described species of mosquitoes in three subfamilies: Toxorhynchitinae, Anophelinae, and Culicinae (the largest). Several closely related families of Nematocera may be confused with the Culicidae owing to a superficial resemblance of adults and larvae. These are the Chironomidae, Dixidae, and Chaoboridae (see Exercise 11). Among them the chaoborids are of the most interest in that adults of a few species are hematophagous and act as vectors of trypanosomes of cold-blooded vertebrates. For this reason keys to North American genera of adult (Exercise 11) and larval (Exercise 12) chaoborids are included in the manual.

These keys to Culicidae include all genera of the New World for adult females and larvae. Males may be distinguished from females by the bushy antennae of the former and by the prominent claspers of the male genitalia (see Figure 12.6). Both sexes of *Anopheles* have palpi subequal in length to the proboscis (Figures 12.3, 12.5), but females of other genera have short palpi (Figures 12.11, 12.12).

Key to mosquito genera of the New World, adult females[a]

---

1. Proboscis strongly recurved; posterior edge of wing emarginated just beyond tip of vein Cu_2 (Figure 12.8) ....*Toxorhynchites*
   - Proboscis not recurved and posterior edge of wing straight or rounded (Figures 12.4, 12.5) 2

2. Palpi about as long as proboscis (Figure 12.5); abdomen with sterna and usually terga wholly or largely devoid of scales 3
   - Palpi much shorter than proboscis (Figure 12.4); abdomen with terga and sterna densely and uniformly covered with scales 4

3. Scutellum trilobed, with setae in 3 distinct groups; 2 prominent lateral tufts of long spatulate scales just anterior to wing bases ....*Chagasia*
   - Scutellum evenly rounded and setae evenly distributed (Figure 12.15); without tufts of spatulate scales anterior to wing bases ....*Anopheles*

4. Cell R_2 of wing always smaller than vein R_{2+3}; apex of anal veins ending before fork of veins Cu_1 and Cu_2; thorax usually with lines of bluish scales ....*Uranotaenia*
   - Cell R_2 at least as long as vein R_{2+3} (Figure 12.7); or if not (*Haemagogus*), then apex of anal vein ends distal to fork of Cu_1 and Cu_2; thorax without lines of bluish scales 5

5. Mesopostnotum always with long setae (Figure 12.10), and sometimes with scales; base of hind coxa usually in line with base of mesomeron or slightly above it; mesomeron very small 6
   - Mesopostnotum without setae or with 2 very small ones in posterior region (Figure 12.1); base of hind coxa distinctly ventral to a larger mesomeron 13

6. Prespiracular area with broad scales only, without setae; hind tarsus with only 1 claw . . . .*Limatus*
   - Prespiracular area with 1 or more setae; hind tarsus with 2 claws 7

7. Antenna at most 0.5 length of proboscis; proboscis 1.16–1.25 times longer than forefemur . . . .*Phoniomyia*
   - Antenna more than 0.5 length of, usually subequal to, proboscis; proboscis shorter or longer than forefemur 8

8. Antepronotal lobes small, well separated; occiput with line of conspicuous dark erect scales posteriorly 9
   - Antepronotal lobes large and approaching at middorsal line; occiput without line of dark erect scales posteriorly 12

9. Proboscis 0.85–1.20 times length of forefemur; row of lower mesokatepisternal setae extending dorsad to above level of ventral border of mesanepimeron 10
   - Proboscis 1.20–1.40 times length of forefemur; row of lower mesokatepisternal setae usually not extending dorsad to level of ventral border of mesanepimeron 11

10. Hind tibia without postmedian light-scaled band; laterotergite of abdominal tergum I without scales basad and sparsely scaled distad, its lower margin visible; postprocoxal membrane without scales . . . .*Trichoprosopon*
    - Hind tibia with broad complete or incomplete postmedian light-scaled band; laterotergite densely scaled, its lower margin hidden; postprocoxal membrane with scales . . . .*Shannoniana*

11. Mid and hind tarsi partly light-scaled; scutal scales moderately broad and flat; scales of vertex and occiput with brilliant silver and azure blue reflections . . . .*Johnbelkinia*
    - Mid and hind tarsi dark-scaled; scutal scales narrow and curved or scales of vertex and occiput without silver reflections and with only weak to moderate green or blue reflections . . . .*Runchomyia*

12. Posterior mesanepisternal setae absent; scutum covered with flat metallic scales with bright irridescent reflections; often with tibio-tarsal "paddles" of erect scales on hind legs . . . .*Sabethes*
    - Posterior mesanepisternal setae present; scutum usually dark-colored, without metallic scales; without "paddles" on legs . . . .*Wyeomyia*

13. Scutum without visible setae on disc; scales smooth and with metallic color; antepronotal lobes enlarged and approaching at middorsal line . . . .*Haemagogus*
    - Scutum with at least prescutellar setae well developed; scutal scalation various but not smooth and metallic in appearance; antepronotal lobes small 14

14. Postpiracular setae present (Figure 12.2) 15
    - Postspiracular setae absent 18

15. Plume scales on dorsal surface of wing veins R_2 and R_3 very broad (Figure 12.13); apex of abdomen blunt 16
    - Plume scales on dorsal surface of wing veins R_2 and R_3 narrow; apex of abdomen more or less pointed 17

16. Femora with conspicuous preapical pale-scaled band; wings
    dark-scaled . . . .*Coquillettidia*
  - Femora marked with dark and pale scales but without definite
    preapical pale band; wing speckled with dark and pale scales . . . . *Mansonia*

17. Prespiracular setae present; bands or patches of pale scales
    apical in position on abdominal terga . . . . *Psorophora*
  - Prespiracular setae absent; bands or patches of pale scales basal
    in position on abdominal terga . . . . *Aedes*

18. Prespiracular setae present; base of subcostal wing vein with
    row of setae on ventral side . . . . *Culiseta*
  - Prespiracular setae absent; base of subcosta without row of
    setae on ventral side 19

19. Tarsomere 4 on fore- and midlegs short, no longer than wide
    (Figure 12.14); scutum usually with fine lines of pale scales . . . .*Orthopodomyia*
  - Tarsomere 4 on fore- and midlegs longer than wide; scutum
    without fine lines of pale scales, with other patterns 20

20. Antenna with short and thick flagellomeres (Figure 12.9);
    midfemur with a tuft of long scales; pulvilli absent . . . .*Aedeomyia*
  - Antenna with elongate flagellomeres; midfemur without a tuft
    of long scales; pulvilli present 21

21. Antenna about as long as proboscis; first flagellomere of
    antenna not much longer than second, if at all . . . .*Culex*
  - Antenna much longer than proboscis, flagellomere 1 definitely
    longer than flagellomere 2 22

22. Flagellomere 1 about 1.5 times as long as flagellomere 2;
    proboscis swollen apically (Figure 12.12) . . . .*Galindomyia*
  - Flagellomere 1 about 2.5 times as long as flagellomere 2;
    proboscis not swollen apically (Figure 12.11) . . . .*Deinocerites*

[a]Prepared by Richard F. Darsie, Jr., and Stephanie Clark Gil.

Key to mosquito genera of the New World, pupae[a]

1. Setae 9-III–VII stout, spinelike, and placed at posterior lateral
   corners of the abdominal segments (Figure 12.22)
   (Anophelinae) 2
  - Setae 9-III–VII usually minute to small, hairlike, placed
    somewhat anterior to the posterior lateral corners of the
    abdominal segments 3

2. Seta 2-III–VII a short, stout, dark spine . . . .*Chagasia*
  - Seta 2-III–VII thin, usually multibranched (Figure 12.22) . . . .*Anopheles*

3. Genital pouch bearing a pair of setae; paddle with posterior
   prolongation of that portion lateral to the midrib (Figure
   12.18) . . . .*Toxorhynchites*
  - Genital pouch without setae; paddle without posterior lateral
    prolongation 4

4. Paddle without true setae, although spinules may be present,
   more or less pointed apically; seta 1 on abdominal segment IX
   absent 5

- Paddle with seta 1 and sometimes seta 2 present, rounded in shape apically; seta 1 on abdominal segment IX present or absent 14

5. Respiratory trumpet attenuated distally, fitted for piercing plant tissue (Figure 12.16); paddle narrow, notched along posterior border, pointed apically 6
- Respiratory trumpet without attenuation distally (Figure 12.20); paddle narrowing apically, but not pointed 7

6. Setae 1 and 5 on abdominal segments III–VII long and stout; seta 6-I–VI present . . . . *Mansonia*
- Setae 1 and 5 on III–VII short and thin; seta 6-I–VI absent . . . . *Coquillettidia*

7. Paddles short, and broad at base, shorter than seta 9-VIII . . . . *Limatus*
- Paddles longer, not so broad at base, usually longer than seta 9-VIII 8

8. Paddles with fringe of long setae on external and internal margins (Figure 12.17) or genital sac borne in deep indenture in segment VIII . . . . *Wyeomyia* (in part)
- Paddle without fringe of large setae, at most fringe at apex; genital sac not borne in deep indenture in VIII 9

9. Trumpet slender, tubular, length/greatest width 9.0 or more (Figure 12.19); paddles broad to near apex, apex pointed . . . . *Phoniomyia*
- Trumpet broader, rather conical, length/greatest width usually no more than 5.0; paddles conical-shaped 10

10. Seta 5-IV–VI long, stout, much longer than succeeding segment, usually single; setae 5-II–III usually weakly developed, not both long; paddle conical or evenly rounded . . . . *Sabethes, Wyeomyia* (in part)
- Seta 5-IV–VI weak, or if long and stout, then with 2 branches and seta 5-II, III long and well developed; paddle usually pear-shaped with lateral and medial indentures near apex 11

11. Seta 5-III–VI weakly developed and short . . . . *Trichoprosopon*
- Seta 5-IV and usually 5-III, V or 5-V, VI moderately to strongly developed and moderately long to long 12

12. Seta 5-C short and weakly developed; seta 3-III at same level or cephalad of 4-III, not close to posterior margin of segment . . . . *Shannoniana*
- Seta 5-C moderately to strongly developed and moderately long to long; seta 3-III caudad of 4-III, near posterior margin 13

13. Part of paddle medial to midrib aculeate; seta 5-II, III moderate to strong, long; 6-VII moderately to strongly developed with 3 or more branches, lateral and anterior to 9-VII . . . . *Johnbelkinia*
- Part of paddle medial to midrib glabrous; seta 5-II, III weak, short, or 6-VII weakly developed single or double, medial to 9-VII . . . . *Runchomyia*

14. Respiratory trumpet with base ringed with tracheoid annulus, usually including at least basal 0.25 (Figure 12.20) 15
- Respiratory trumpet mostly with reticular surface, if tracheoid patch present basally, not completely encircling basal portion 19

15. Paddle widest in apical 0.25, with deep cleft apically . . . . *Aedeomyia*
    • Paddle widest in basal 0.66, without deep cleft apically 16

16. Paddle conspicuously unequally divided by midrib, larger and
    lobelike medially, with prominent spicules on lateral and
    medial margins, seta 1-IX usually a well developed seta (Figure
    12.27) . . . . *Uranotaenia*
    • Paddle subequally divided by midrib, margins smooth or with
    tiny spicules (Figure 12.23); seta 1–IX very tiny or absent 17

17. Seta 9-VIII single, longer than length of VIII; paddle seta 1
    0.66 or more length of paddle, paddle seta 2 absent (Figure
    12.24) . . . . *Deinocerites*
    • Seta 9-VIII usually not single, or if so, then shorter than length
    of VIII; paddle seta 1 no more than 0.5 length of paddle,
    paddle seta 2 present or absent 18

18. Seta 9-II–VI attached at caudolateral angle; seta 6-VII basad
    of 9-VII; seta 2-VI laterad of 1-VI . . . . *Galindomyia*
    • Seta 9-II-VI attached somewhat basad to caudolateral angle
    (Figure 12.21); seta 6-VII caudad of 9-VII; seta 2-VI mediad
    to 1-VI . . . . *Culex*

19. Abdominal segment VIII not markedly smaller basally, so that
    VIII and VII appear fused; paddle seta 1 multibranched . . . . *Orthopodomyia*
    • Abdominal segment VIII smaller basally so that VIII and VII
    appear as distinctly separate segments; paddle seta 1 usually
    single or double 20

20. Seta 5-II, III short, weak, single, or if not (subgenus
    Conopostegus), paddle seta 1 more than 0.5 length of paddle;
    seta 11-T longer and stronger than seta 10-T; midrib usually
    strongly sclerotized, reaching to apex . . . . *Haemagogus*
    • Seta 5-II, III moderately to strongly developed, or if not,
    multibranched on both segments; paddle seta 1 short, less than
    0.5 length of paddle (Figure 12.25); seta 11-T usually not
    longer or stronger than 10-T; paddle midrib not strongly
    sclerotized, nor reaching apex 21

21. Seta CT-8 and CT-9 in perpendicular line to middorsal ridge . . . . *Culiseta*
    • Seta CT-8 usually distinctly more anterior than CT-9 (see
    Figure 12.20) 22

22. Seta 6 on abdominal segment VII anterior to or mesad of seta
    9; seta 6 on abdominal segment I with 2 or more branches,
    rarely single; paddle with seta 2 present and/or with sublateral
    ventral lobe along posterior border of VIII (Figure 12.26) . . . . *Psorophora*
    • Seta 6-VII distinctly posterior to 9-VII; 6-I single rarely
    double; paddle seta 2 absent; without sublateral, ventral lobe
    along posterior border of VIII . . . . *Aedes*

*a*Prepared by Richard F. Darsie, Jr.

Key to mosquito genera of the New World, fourth-stage larvae[a]

1. Abdominal segment VIII without respiratory siphon; some
   abdominal segments with palmate setae (Figure 13.3) 2
   • Abdominal segment VIII with a siphon; abdominal segments
   without palmate setae (Figure 12.28) 3

2. Palmate setae racket-shaped (Figure 12.31), present on abdominal segments III–V; anterior lobe of spiracular apparatus produced into a long process, ending in an equally long seta . . . .*Chagasia*

• Palmate setae spindle-shaped (Figure 13.3), sometimes present on abdominal segments in position 1; anterior lobe of spiracular apparatus normal . . . .*Anopheles*

3. Setae 4-X composed of 1 pair of setae 4

• Setae 4-X forming a ventral brush, composed of at least 4 pairs of setae (Figure 12.30) 11

4. Maxilla with a large horn, modified for grasping (Figure 12.33) 5

• Maxilla normal, not modified for grasping (Figure 12.29) 7

5. Head capsule with short to long transverse slitlike foramen magnum, not bounded by distinct collar; siphon with filamentous pecten . . . .*Runchomyia*

• Head capsule with normal circular foramen magnum bounded by distinct collar; siphon without pecten 6

6. Siphon without accessory midventral setae; maxilla with strong articulation to head capsule far ventrad of palpus; seta 8-M weakly developed . . . .*Shannoniana*

• Siphon with long row of multibranched accessory midventral setae; maxilla without strong articulation to head capsule far ventrad of palpus; seta 8-M strongly developed (see Figure 12.28) . . . .*Johnbelkinia*

7. Mandible enlarged, seta 8-M absent . . . .*Trichoprosopon*

• Mandible normal, seta 8-M present 8

8. Siphon less than 3 times as long as greatest width (Figure 12.32); siphonal setae mostly branched . . . .*Limatus*

• Siphon more than 4 times as long as greatest width; siphonal setae mostly single 9

9. Seta 4 as long as seta 3 on anal segment . . . .*Sabethes*

• Seta 4 shorter than seta 3 on anal segment 10

10. Siphon more or less acutely attenuated apically, index usually 6.0 or greater (Figure 12.34) . . . .*Phoniomyia*

• Siphon not acutely attenuated apically, more blunt, index usually less than 6.0 . . . .*Wyeomyia*

11. Siphon short, attenuated apically, fitted for piercing plants (Figure 12.35) 12

• Siphon more or less cylindrical, not attenuated apically 13

12. Saddle of anal segment without long setae ventrally (Figure 12.35); seta 2-, 3-A short, part of antenna distal to their point of attachment much longer than part basal to it . . . .*Coquillettidia*

• Saddle with 3–4 long setae ventrally (Figure 12.36); seta 2-, 3-A long, as long as part of antenna distal to their point of attachment, this part no longer than basal part . . . .*Mansonia*

13. Siphon without pecten 14

• Siphon with pecten 16

14. Dorsal and ventral abdominal setae in groups of 3–5 on large common sclerotized plates; lateral palatal brushes reduced to

about 12 broad flat simple filaments (Figure 12.40); comb
scales absent on VIII . . . . *Toxorhynchites*
- Dorsal and ventral abdominal setae arising separately and
  without strong basal plates; lateral palatal brushes of at least 40
  thin cylindrical simple or pectinate filaments; comb scales
  present on VIII 15

15. Antenna simple, much shorter than head capsule; surface of
    siphon glabrous . . . . *Orthopodomyia*
- Antenna strongly curved and longer than head capsule (Figure
  12.37); surface of siphon pilose . . . . *Aedeomyia*

16. Comb scales arising from large sclerotized plate on abdominal
    segment VIII (Figure 12.38); head longer than wide . . . . *Uranotaenia*
- Comb scales usually not attached to a plate, or if so, plate very
  small; head wider than long 17

17. Siphon with only one pair of setae in addition to setae 2-S
    (Figure 12.30) 18
- Siphon with 3 or more pairs of setae in addition to 2-S 21

18. Ventral brush of anal segment usually with at least 4 precratal
    setae attached to complete saddle, or if saddle incomplete,
    precratal setae extending to basal 0.5 of segment (see Figure 12.36) . . . .*Psorophora*
- Ventral brush without precratal setae attached to saddle when
  complete, or when incomplete, precratal setae confined to
  apical 0.33 of segment 19

19. Anal saddle complete (Figure 12.30) . . . . *Aedes* (in part)
- Anal saddle incomplete 20

20. Seta 3-VII stout and long; posterior margin of saddle with long
    spines . . . . *Haemagogus*
- Seta 3-VII medium to small and short; posterior margin of
  saddle with or without short spines . . . . *Aedes* (in part)

21. Siphon with a pair of setae near base (Figure 12.39); pecten
    usually followed by row of closely spaced single setae (may be
    setal tufts) . . . .*Culiseta*
- Siphon without pair of setae near base; pecten followed by
  various setae, not closely set row 22

22. Comb scales numbering 16 or fewer; saddle sclerite entire, not
    encircling anal segment (Figure 12.28) . . . . *Aedes* (in part)
- Comb scales numbering more than 16; saddle sclerite entire,
  completely encircling anal segment, or divided into dorsal and
  ventral plates 23

23. Seta 1-VI mesad of seta 2-VI; pecten spine untoothed; setae
    2,3-C absent . . . . *Galindomyia*
- Seta 1-VI laterad of seta 2-VI; pecten spines usually with one
  or more lateral teeth; setae 3-C and often 2-C present
  (see Figure 12.29) 24

24. Head capsule widest near level of antennal bases (Figure
    12.41); two small sclerites present on anal segment . . . . *Deinocerites*
- Head capsule widest in caudal half; one large sclerite present
  on anal segment, usually completely encircling segment . . . . *Culex*

---

[a]Prepared by Richard F. Darsie, Jr. and Stephanie Clark Gil.

Key to genera of Chaoboridae of North America, larvae

1. Eighth abdominal segment with no siphon or external trace of respiratory apparatus (Figure 12.42) ....*Chaoborus*
   - Eighth abdominal segment with a siphon or at least a prominent spiracular apparatus 2

2. Antennae approximate (close together); each ocular lobe with a transverse row of stout setae (Figure 12.43) ....*Corethrella*
   - Antennae widely separated; no row of stout setae on ocular lobes 3

3. Eighth abdominal segment with an elongate siphon (Figure 12.44) ....*Mochlonyx*
   - Eighth abdominal segment with a short spiracular apparatus similar to *Anopheles* (Figure 12.45) ....*Eucorethra*

Genera of Culicidae recognized in the New World

1. *Aedeomyia* Theobald
2. *Aedes* Meigen
3. *Anopheles* Meigen
4. *Chagasia* Cruz
5. *Coquillettidia* Dyar
6. *Culex* Linnaeus
7. *Culiseta* Felt
8. *Deinocerites* Theobald
9. *Galindomyia* Stone & Barreto
10. *Haemagogus* Williston
11. *Johnbelkinia* Zavortink
12. *Limatus* Theobald
13. *Mansonia* Blanchard
14. *Orthopodomyia* Theobald
15. *Phoniomyia* Theobald
16. *Psorophora* Robineau-Desvoidy
17. *Runchomyia* Theobald
18. *Sabethes* Robineau-Desvoidy
19. *Shannoniana* Lane & Cerquiera
20. *Toxorhynchites* Theobald
21. *Trichoprosopon* Theobald
22. *Uranotaenia* Lynch Arribalzaga
23. *Wyeomyia* Theobald

References

Adults and larvae

Adames, A. J. 1971. Mosquito studies (Diptera: Culicidae). 24. A revision of the crabhole mosquitoes of the genus *Deinocerites*. *Contr. Am. Entomol. Inst. 7*(2):1–154.

Arnell, J. H. 1976. Mosquito studies (Diptera: Culicidae). 33. A revision of the *scapularis* group of *Aedes* (*Ochlerotatus*). *Contr. Am. Entomol. Inst. 13*(3):1–144.

Belkin, J. N., S. J. Heinemann, and W. A. Page. 1970. Mosquito studies (Diptera: Culicidae) XXI. The Culicidae of Jamaica. *Contr. Am. Entomol. Inst. 6*(1):1–458.

Bohart, R. M., and R. K. Washino. 1978. *Mosquitoes of California*, 3rd ed. Publ. 4084. University of California, Division of Agricultural Sciences, Berkeley. 153 pp.

Bram, R. A. 1967. Classification of *Culex* subgenus *Culex* in the New World (Diptera: Culicidae). *Proc. U.S. Nat. Mus. 120* (3557), 122 pp.

Carpenter, S. J., and W. J. La Casse. 1955. *Mosquitoes of North America* (*North of Mexico*). University of California Press, Berkeley. 360 pp.

Forattini, O. P. 1962. *Entomología médica: Parte geral, Diptera, Anophelini*, vol. I. University of São Paulo. 662 pp.

– 1965. *Entomología médica. Culicini: Culex, Aedes e Psorophora*. vol. II, University of São Paulo, 506 pp.

Foote, R. H. 1954. *The larvae and pupae of the mosquitoes belonging to the Culex subgenera Melanoconion and Mochlostyrax*. USDA Tech. Bull. 1091, 125 pp.

Harbach, R. E., and K. L. Knight. 1980. *Taxonomist's Glossary of Mosquito Anatomy*. Plexus, Marlton, N.J. 415 pp.

Knight K. 1978. *Supplement to a Catalog of the Mosquitoes of the World* (*Diptera: Culicidae*), suppl. to vol. 6. Thomas Say Foundation, College Park, Md. 107 pp.

Knight, K., and A. Stone. 1973. A Catalog of the Mosquitoes of the World (*Diptera: Culicidae*), vol. 6, 2nd ed. Thomas Say Foundation, College Park, Md. 611 pp.

Lane, J. 1953. *Neotropical Culicidae*, vol. 1/2. University of São Paulo. 1112 pp.

Mattingly, P. 1971. Contributions to the mosquito fauna of Southeast Asia XII. Illustrated keys to the genera of mosquitoes (Diptera: Culicidae). *Contr. Am. Entomol. Inst. 7*(4):1–84.

Zavortink, T. J. 1968. Mosquito studies (Diptera: Culicidae) VIII. A prodrome of the genus *Orthopodomyia*. *Contr. Am. Entomol. Inst. 3*(2):1–221.

– 1969. Mosquito studies (Diptera, Culicidae) XIX. The treehole *Anopheles* of the New World. *Contr. Am. Entomol. Inst. 5*(2):1–35.

– 1972. Mosquito studies (Diptera: Culicidae) XXVIII. The New World species formerly placed in *Aedes* (*Finlaya*). *Contr. Am. Entomol. Inst. 8*(3):1–206.

– 1979. Mosquito studies (Diptera: Culicidae) XXXV. The new sabethine genus *Johnbelkinia* and a preliminary reclassifica-

tion of the composite genus *Trichoprosopon*. *Contr. Am. Entomol. Inst. 17*(1):1–61.
- 1979. A reclassification of the sabethine genus *Trichoprosopon Mosq. Syst. 11*(4):255–7.

Pupae

Barr, A. R. 1963. Pupae of the genus *Culiseta* Felt II. Descriptions and a key to the North American species (Diptera: Culicidae). *Ann. Entomol. Soc. Am. 56*(3):324–30.
Barr, A. R., and S. Barr. 1969. Mosquito studies (Diptera, Culicidae) XIII. Pupae of the genus *Psorophora* in North America and Puerto Rico. *Contr. Am. Entomol. Inst. 4*(4):1–20.
Belkin, J. N. 1950. A revised nomenclature for the chaetotaxy of the mosquito larva (Diptera: Culicidae). *Am. Midl. Nat. 44*:678–98.

- 1952. The homology of the chaetotaxy of immature mosquitoes and a revised nomenclature for the chaetotaxy of the pupa (Diptera, Culicidae). *Proc. Entomol. Soc. Wash. 54*:115–30.
Darsie, R. F., Jr. 1949. Pupae of the anopheline mosquitoes of the Northeastern United States (Diptera, Culicidae). *Rev. Entomol. 20*(1):509–30.
- 1951. Pupae of the culicine mosquitoes of the Northeastern United States (Diptera, Culicidae, Culicini). Cornell Univ. Agr. Exp. Sta. Mem. 304. Cornell University, Ithaca, N.Y. 67 pp.
Knight, K. L. 1971. A mosquito taxonomic glossary VII. The pupa. *Mosq. Syst. Newsl. 3*:42–65.
Mattingly, P. 1971. Contributions to the mosquito fauna of Southeast Asia XII. Illustrated keys to the genera of mosquitoes (Diptera: Culicidae). *Contr. Am. Entomol. Inst. 7*(4):1–84.

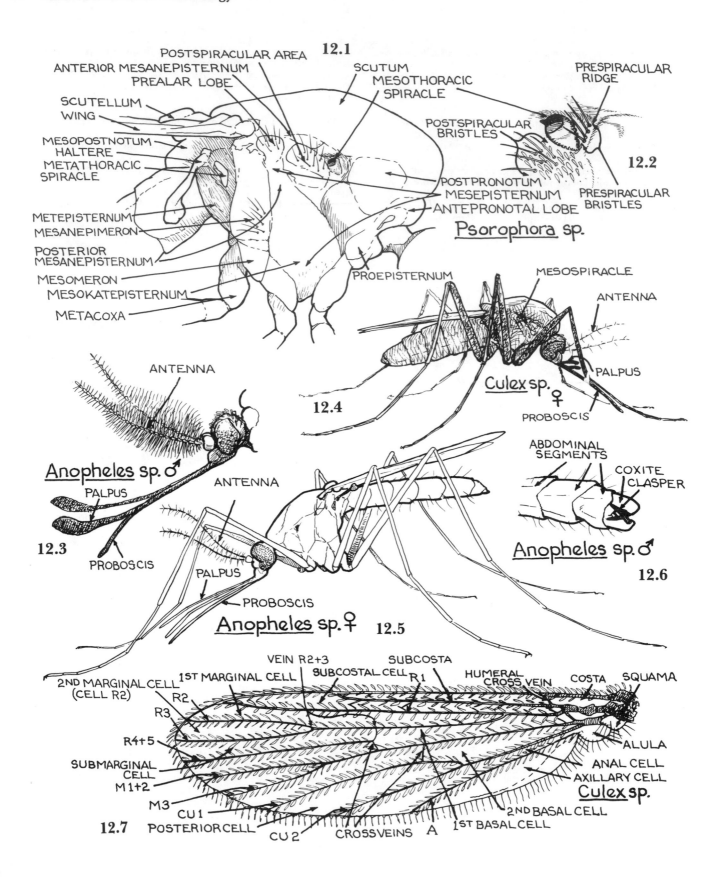

12.1

POSTSPIRACULAR AREA
ANTERIOR MESANEPISTERNUM
PREALAR LOBE
SCUTELLUM
WING
MESOPOSTNOTUM
HALTERE
METATHORACIC
SPIRACLE
METEPISTERNUM
MESANEPIMERON
POSTERIOR
MESANEPISTERNUM
MESOMERON
MESOKATEPISTERNUM
METACOXA
PROEPISTERNUM
SCUTUM
MESOTHORACIC
SPIRACLE
POSTPRONOTUM
MESEPISTERNUM
ANTEPRONOTAL LOBE

12.2
PRESPIRACULAR
RIDGE
POSTSPIRACULAR
BRISTLES
PRESPIRACULAR
BRISTLES

Psorophora sp.

12.3
ANTENNA
Anopheles sp. ♂
PALPUS
PROBOSCIS

12.4
MESOSPIRACLE
ANTENNA
PALPUS
Culex sp. ♀
PROBOSCIS

12.5
ANTENNA
PALPUS
PROBOSCIS
Anopheles sp. ♀

12.6
ABDOMINAL
SEGMENTS
COXITE
CLASPER
Anopheles sp. ♂

12.7
2ND MARGINAL CELL
(CELL R2)
1ST MARGINAL CELL
VEIN R2+3
SUBCOSTAL CELL
SUBCOSTA
R2
R1
R3
HUMERAL
CROSS VEIN
COSTA
SQUAMA
R4+5
SUBMARGINAL
CELL
M1+2
M3
ALULA
ANAL CELL
AXILLARY CELL
Culex sp.
CU1
POSTERIOR CELL
CU2
CROSSVEINS
A
1ST BASAL CELL
2ND BASAL CELL

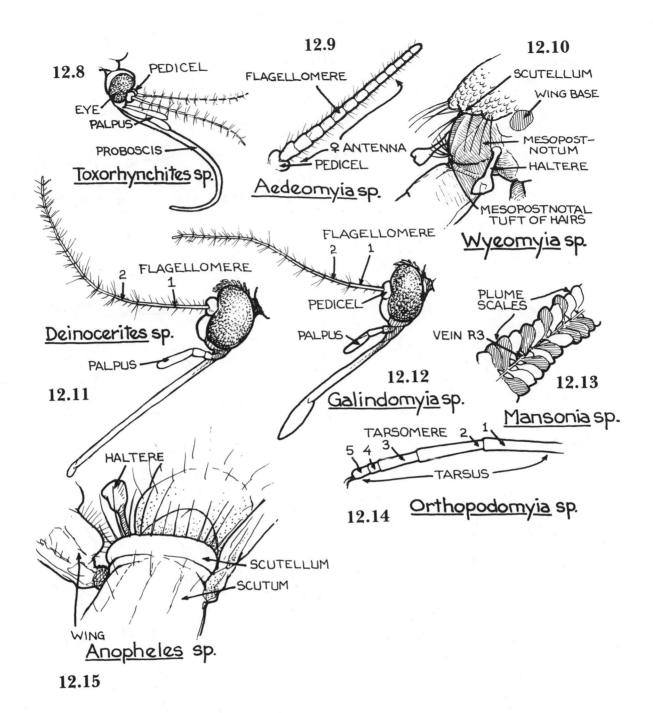

12.8 Toxorhynchites sp.

PEDICEL
EYE
PALPUS
PROBOSCIS

12.9 Aedeomyia sp.

FLAGELLOMERE
♀ ANTENNA
PEDICEL
♂

12.10 Wyeomyia sp.

SCUTELLUM
WING BASE
MESOPOST-NOTUM
HALTERE
MESOPOSTNOTAL TUFT OF HAIRS

12.11 Deinocerites sp.

FLAGELLOMERE
2
1
PALPUS

12.12 Galindomyia sp.

FLAGELLOMERE
2
1
PEDICEL
PALPUS

12.13 Mansonia sp.

PLUME SCALES
VEIN R3

12.14 Orthopodomyia sp.

TARSOMERE
5 4 3 2 1
TARSUS

12.15 Anopheles sp.

HALTERE
SCUTELLUM
SCUTUM
WING

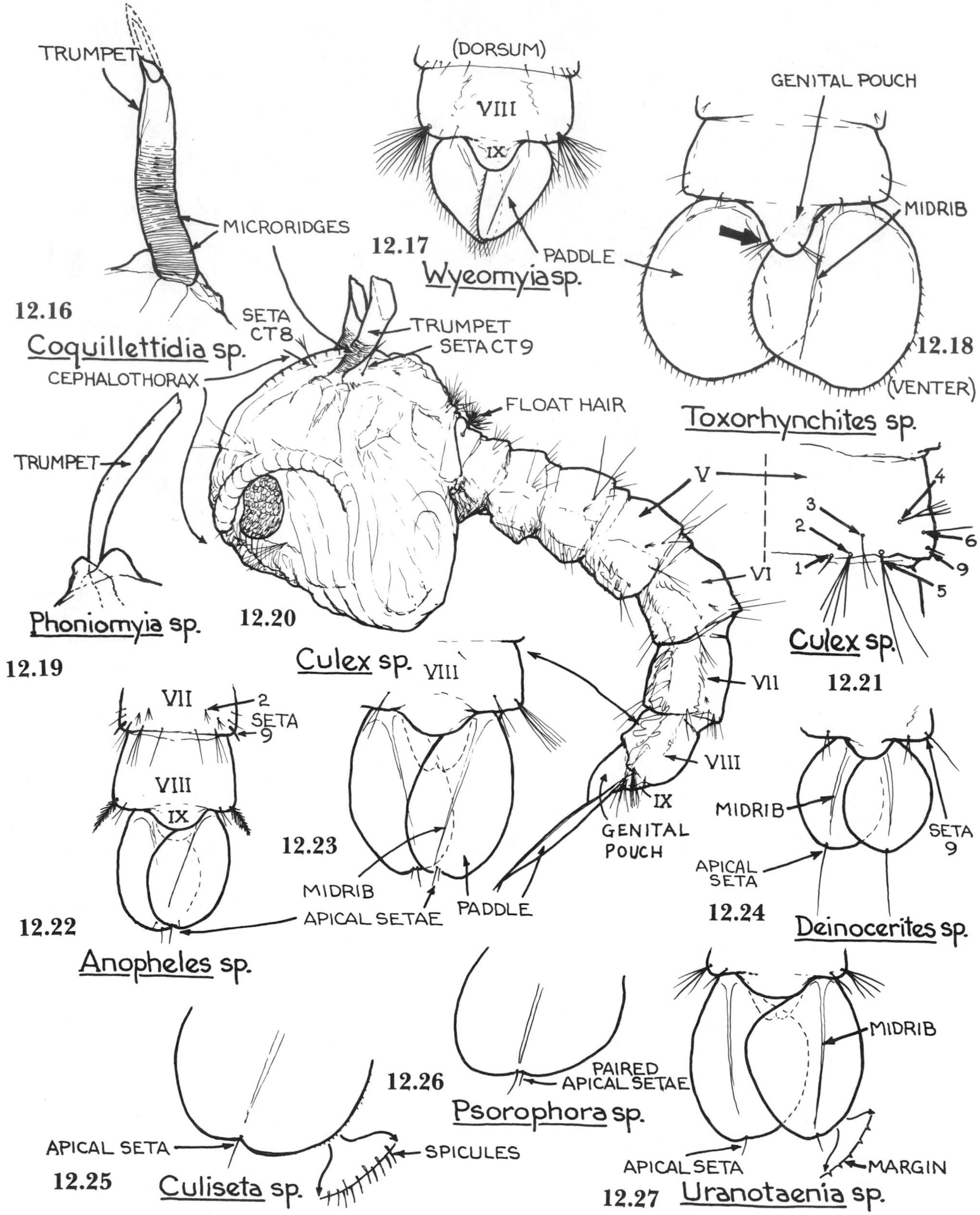

12.16
Coquillettidia sp.
TRUMPET
MICRORIDGES

12.17
Wyeomyia sp.
(DORSUM)
VIII
IX
PADDLE

12.18
(VENTER)
Toxorhynchites sp.
GENITAL POUCH
MIDRIB

12.19
Phoniomyia sp.
TRUMPET

12.20
Culex sp.
SETA CT8
CEPHALOTHORAX
TRUMPET
SETA CT9
FLOAT HAIR
V
VI
VII
VIII
IX
GENITAL POUCH

12.21
Culex sp.
1 2 3 4 5 6 9

12.22
Anopheles sp.
VII
SETA 9
VIII
IX

12.23
MIDRIB
APICAL SETAE
PADDLE
VIII

12.24
Deinocerites sp.
MIDRIB
APICAL SETA
SETA 9

12.25
Culiseta sp.
APICAL SETA
SPICULES

12.26
Psorophora sp.
PAIRED APICAL SETAE

12.27
Uranotaenia sp.
MIDRIB
APICAL SETA
MARGIN

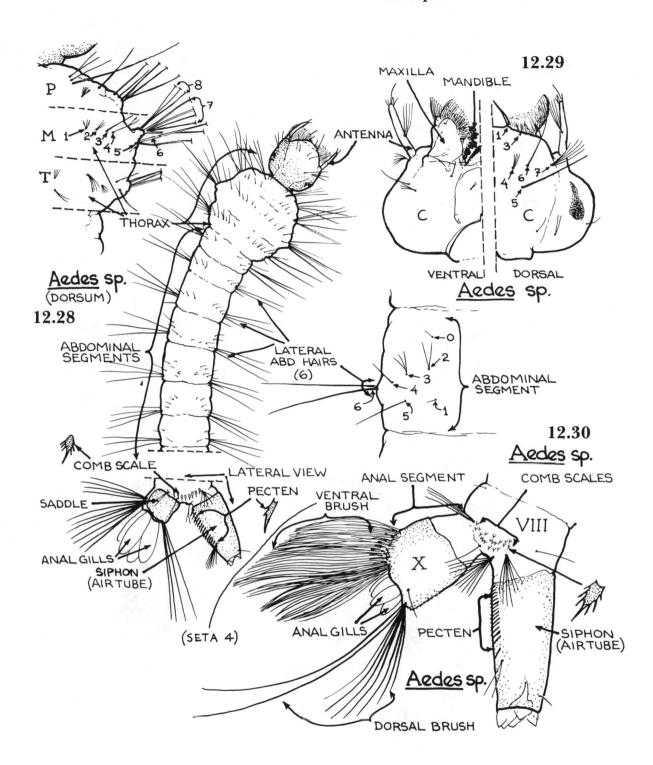

12.28

Aedes sp. (DORSUM)

P
M 1 2 3 4 5 6 7 8
T
THORAX
ANTENNA

ABDOMINAL SEGMENTS
LATERAL ABD HAIRS (6)

COMB SCALE

SADDLE
PECTEN
ANAL GILLS
SIPHON (AIRTUBE)
LATERAL VIEW
(SETA 4)

12.29
MAXILLA MANDIBLE

VENTRAL   DORSAL
C   C
1 3 4 6 7 5

Aedes sp.

0
2
3
4
6
5
1
ABDOMINAL SEGMENT

12.30
Aedes sp.

ANAL SEGMENT
VENTRAL BRUSH
COMB SCALES
VIII
X
ANAL GILLS
PECTEN
SIPHON (AIRTUBE)
DORSAL BRUSH
Aedes sp.

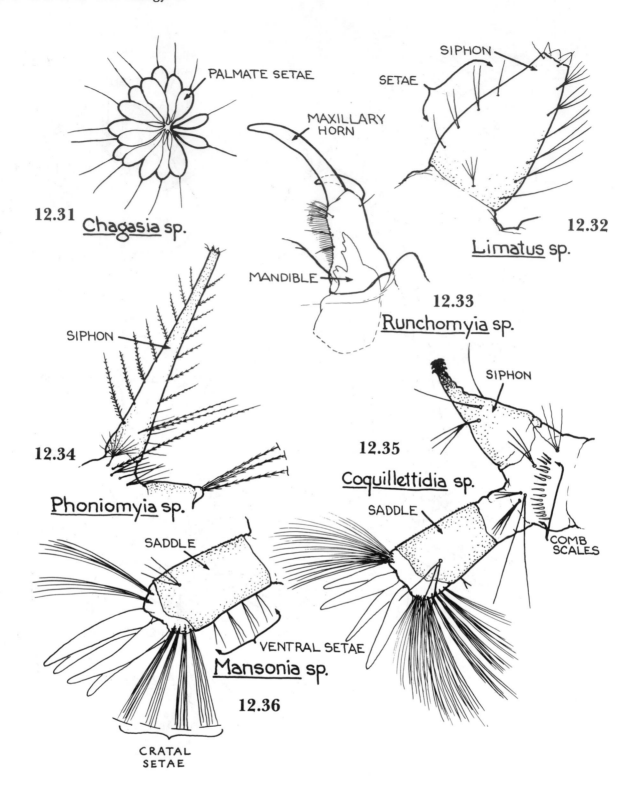

PALMATE SETAE

12.31 <u>Chagasia</u> sp.

SIPHON

SETAE

MAXILLARY HORN

MANDIBLE

12.32 <u>Limatus</u> sp.

12.33 <u>Runchomyia</u> sp.

SIPHON

12.34 <u>Phoniomyia</u> sp.

SIPHON

12.35 <u>Coquillettidia</u> sp.

SADDLE

COMB SCALES

SADDLE

VENTRAL SETAE

<u>Mansonia</u> sp.

12.36

CRATAL SETAE

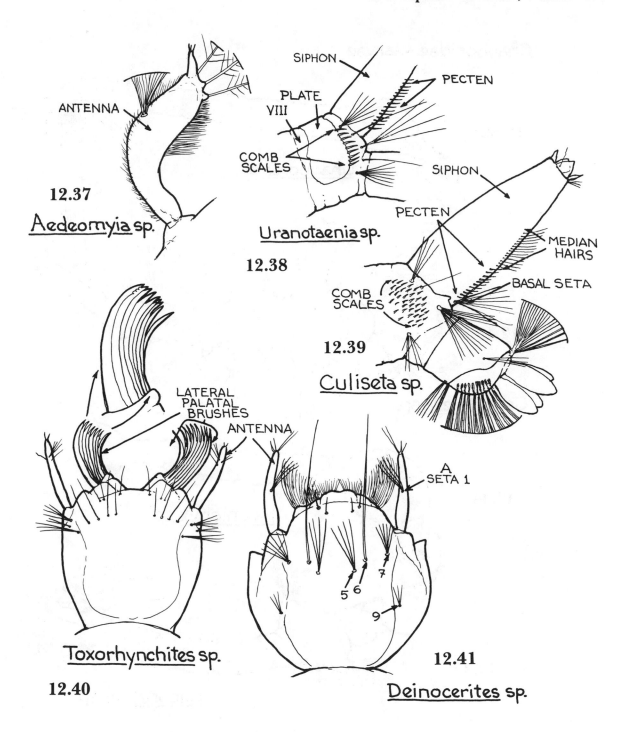

12.37
Aedeomyia sp.

ANTENNA

SIPHON

PECTEN

PLATE
VIII

COMB
SCALES

Uranotaenia sp.

12.38

SIPHON

PECTEN

MEDIAN
HAIRS

COMB
SCALES

BASAL SETA

12.39

Culiseta sp.

LATERAL
PALATAL
BRUSHES

ANTENNA

Toxorhynchites sp.

12.40

A
SETA 1

5 6
7

9

12.41

Deinocerites sp.

Chaoboridae – larvae

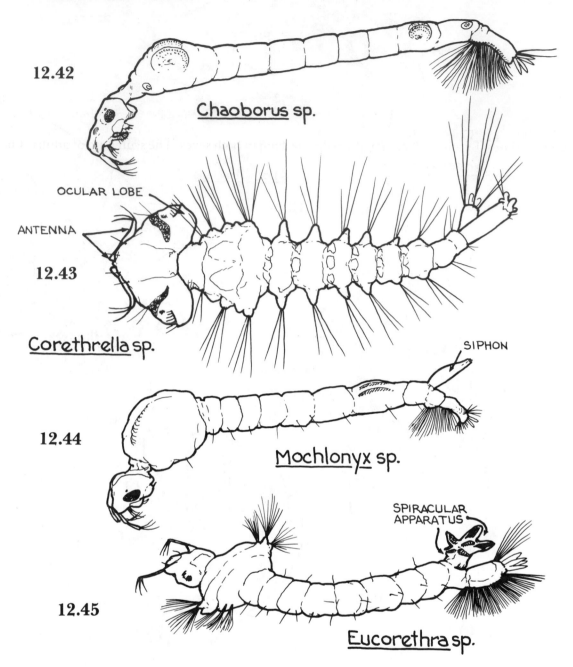

12.42

Chaoborus sp.

OCULAR LOBE

ANTENNA

12.43

Corethrella sp.

SIPHON

12.44

Mochlonyx sp.

SPIRACULAR
APPARATUS

12.45

Eucorethra sp.

13 Order Diptera: *Anopheles*

Anopheles mosquitoes are almost worldwide in distribution. Together with *Bironella* of New Guinea and *Chagasia* of tropical America, *Anopheles* constitute the subfamily Anophelinae of the family Culicidae. In both adult and immature instars the Anophelinae differ in a number of morphological characters from other mosquitoes. Except in *Bironella*, the palpi are long in both sexes. The scutellum of adults of both sexes has an unbroken row of marginal setae and is evenly rounded posteriorly except in *Chagasia*, in which it is slightly trilobed. The abdominal sternites of *Anopheles* adults are essentially without scales. Anopheline larvae lack a dorsal respiratory siphon, or air tube, but the posterodorsal surface of the eighth abdominal segment bears a lateral chitinous plate with a posterior row of teeth, the pecten; additionally there is a series of paired dorsal palmate hairs on all or some of the first seven abdominal segments.

Anopheles species occupy a preeminent position among arthropods of importance to the health of man as *the* vectors of human malaria plasmodia and as important vectors of the *Wuchereria* and *Brugia* filariae causing human filariasis.

Key to species of nearctic *Anopheles*, adults

1. All tarsi dark; wings usually with not more than 2 pale scaled costal areas 2
● Hind tarsus with a broad white band; wings usually with 4 or more pale areas near costal margin (Figure 13.6) ....*albimanus*

2. Wings with areas of light-colored scales 3
● Wings entirely dark-scaled (except for light apical fringe in *earlei* and *occidentalis*) 7

3. Wings with 2 or more costal areas of pale scales; anal vein with 1 to 2 areas of dark scales 5
● Wings with but 1 pale spot at tip of costa; anal vein with 3 areas of dark scales (only 2 in male) 4

4. Gulf and Atlantic coastal region from Honduras to Massachusetts and west to Kansas and New Mexico; Cuba and Jamaica; in semipermanent to permanent pools, lakes, and swamps ....*crucians*
● Atlantic and Gulf coastal regions from New York to Texas, also of Mexico; brackish pools near coast ....*bradleyi*
● Inland southeastern U.S.; in seepage areas and sluggish streams ....*georgianus*

5. Palpi banded with white; wing veins R$_{4-5}$ and cubitus with long areas of pale scales 6
● Palpi unbanded, black; wing veins R$_{4-5}$ and cubitus entirely dark-scaled ....*perplexens; punctipennis*

6. Terminal segment of palpus white-scaled; vein M pale-scaled before the fork ....*pseudopunctipennis*
● Terminal segment of palpus dark-scaled at apex; vein M dark before the fork ....*franciscanus*

7. Wings with more or less distinct dark spots of clumped scales (Figure 13.7); setae on mesonotum rarely half as long as width of mesonotum; mesonotum dull on rubbed specimens 8

- Wings unspotted; setae on mesonotum at least half as long as width of mesonotum; mesonotum shiny on rubbed specimens .... *barberi*

8. Tip of wing with a silver or copper-colored fringe spot 9
- Tip of wing lacking pale fringe spot .... .10

9. Scales on stem of R_{2-3} between fork and dark spot rather closely appressed (restricted to narrow strip of Pacific Coast) .... *occidentalis*
- Scales on stem of R_{2-3} between fork and dark spot raised (northern U.S. and Canada to Alaska) .... *earlei*

10. Wing with 4 distinct dark spots (Figure 13.7); head with some pale setae in frontal tuft; palpi entirely dark .... .11
- Wing usually with indistinct dark spots; frontal tuft entirely dark; palpal segments with or without distinct white apical rings .... .12

11. Generally occurring west of Rocky Mountains; cubital vein before the fork with scales having truncate, serrate tips (Figure 13.8); mesonotum with indistinct pale stripe .... *freeborni*
- Occurring east of the Rockies; cubital vein before the fork with scales predominantly broadly rounded at apices without serrate edges; mesonotum uniformly colored .... *quadrimaculatus*

12. Palpi with narrow white apical segmental rings; knob of haltere usually pale-scaled; knee spots present .... *walkeri*
- Palpi entirely dark or with faint apical segmental white rings; knob of haltere entirely dark-scaled; knee spots reduced or absent .... *atropos*

Key to species of nearctic *Anopheles*, fourth-stage larvae

1. Lateral hair 6 of abdominal segments 1 to 3 plumose (Figure 13.3); frontal head hairs 5–7 large, plumose (Figure 13.4) 2
- Lateral hair 6 of abdominal segments 1 to 6 plumose; frontal head hairs 5–7 reduced and consisting of single unbranched hairs .... *barberi*

2. Palmate hairs on abdominal segment 1 absent, if present on segment 2, they are smaller than those of segments 3 to 6 (Figure 13.3) 3
- Palmate hairs well developed on abdominal segments 1 to 6 .... *albimanus*

3. Head with outer and inner clypeal hairs simple (Figure 13.1) 4
- Outer clypeal hairs plumose 5

4. Posterior spiracular plate produced into 2 long, black, upturned tails .... *pseudopunctipennis*
- Posterior spiracular plate without tails .... *franciscanus*

5. Outer clypeal hairs thickly, dichotomously branched (more than 25 branches) 6
- Outer clypeal hairs sparsely plumose or branched in apical half (5 to 20 branches) .... *atropos*

6. Inner clypeals separated by at least the width of basal tubercle (Figure 13.1) 7
- Inner clypeals less than width of tubercle apart 8

7. Palmate hairs of second segment well developed and pigmented, although smaller than palmate hairs of segment 3 (common malaria mosquito) ....*quadrimaculatus*
   - Palmate hairs undeveloped and unpigmented except on segments 4, 5, and 6 ....*bradleyi* (see also couplet 10)

8. Fourth abdominal segment with hair "0" missing or reduced and single 9
   - Fourth abdominal segment with hair "0" well developed and multiple-branched (Figure 13.2) 12

9. Palmate hairs well developed and pigmented on only abdominal segments 4, 5, and 6 10
   - Palmate hairs well developed on at least 5 abdominal segments 11

10. Hair 2 of 4th segment usually triple (Figure 13.2) ....*georgianus*
    - Hair 2 of 4th abdominal segment usually single ....*bradleyi*

11. Antepalmate hair 2 of abdominal segments 4 and 5 usually single; postclypeal hairs usually with 3 or more branches ....*occidentalis*
    - Antepalmate hair 2 of abdominal segments 4 and 5 generally double but sometimes single or triple; postclypeal hairs bifurcate ....*freeborni; perplexens; punctipennis* (*perplexans* restricted primarily to limestone springs and their streams in southeastern U.S.); in western U.S. *freeborni* generally has the antennal hair inserted at or beyond the basal third of the tube instead of before the basal third as in *punctipennis*)

12. Inner clypeals with sparse minute feathering toward tip; hair 2 of 4th abdominal segment single ....*walkeri*
    - Inner clypeals bare; hair 2 of 4th segment multiple ....*crucians*

References

(To supplement those given in Exercise 12)
Barr, A. R. 1958. *The Mosquitoes of Minnesota (Diptera: Culicidae: Culicinae)*. Univ. Minn. Agr. Exp. Sta. Tech. Bull. 228. 154 pp.
Bellamy, R. E. 1956. An investigation of the taxonomic status of *Anopheles perplexens* Ludlow, 1907. *Ann. Entomol. Soc. Am.* 49:515–29.
Gjullin, C. M., and G. W. Eddy. 1972. *The Mosquitoes of the North-western United States*. USDA Res. Service Tech. Bull. 1447. 111 pp.
Ross, E. S., and H. R. Roberts. 1943. *Mosquito Atlas (Part I): The Nearctic Anopheles*. The American Entomological Society and Academy of Natural Science, Philadelphia. 48 pp.
Ross, H. H., and W. R. Horsfall. 1965. *A Synopsis of the Mosquitoes of Illinois (Diptera, Culicidae)*. Illinois Natural Hist. Survey Bio. Notes No. 52. 50 pp.

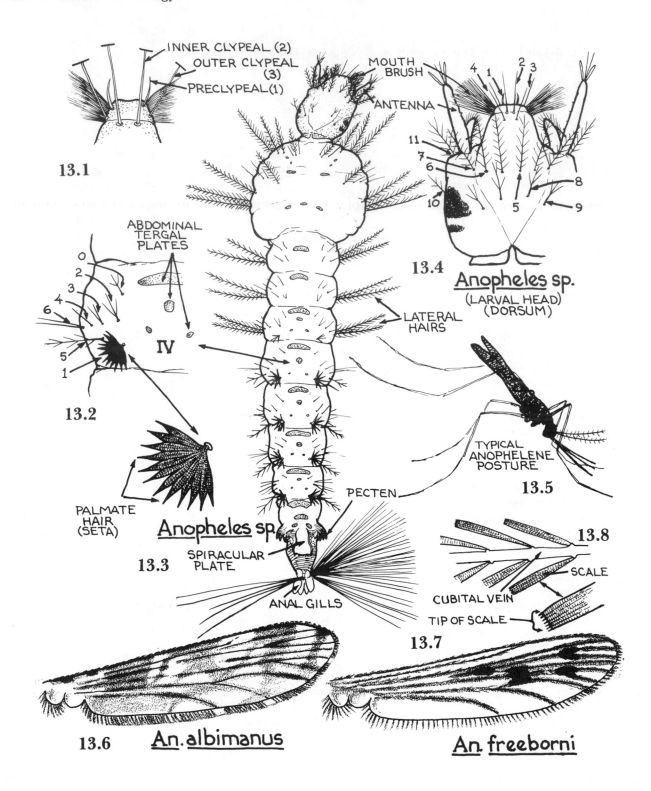

INNER CLYPEAL (2)
OUTER CLYPEAL (3)
PRECLYPEAL (1)

13.1

MOUTH BRUSH
ANTENNA

4 1   2 3
11
7
6
10   5   8
                      9

13.4
Anopheles sp.
(LARVAL HEAD)
(DORSUM)

ABDOMINAL TERGAL PLATES

0
2
3
4
6
5
1

IV

13.2

LATERAL HAIRS

PALMATE HAIR (SETA)

Anopheles sp.

13.3   SPIRACULAR PLATE

TYPICAL ANOPHELENE POSTURE

13.5

PECTEN

ANAL GILLS

13.8

SCALE

CUBITAL VEIN

TIP OF SCALE

13.7

13.6   **An. albimanus**

An. freeborni

14   Order Diptera: Culicinae (adults)

Adult mosquitoes of the subfamily Culicinae are characterized by their trilobed scutellum, which has marginal setae only on the lobes; by broad scales, usually lying flat on the abdomen; and by short palpi in the female. The majority of species in the 20-plus known genera belong to *Aedes* and *Culex*. Many of the Culicinae are important pests of man and animals and serve as vectors for some of our most dangerous pathogens, including the various arboviruses such as Western equine encephalitis, St. Louis encephalitis, and so on, and of human and animal filarial worms, heartworms of dogs, and avian malaria.

Key to commoner species of *Aedes* of North America, adults

1. Tarsal segments, at least on hindlegs, ringed with white
(Figure 14.1)...2
 • Tarsal segments without white rings 16

2. Tarsal segments with white rings on basal but *not* on apical part
of segment 3
 • At least some tarsal segments with white rings both apically and
basally 13

3. Proboscis ringed with white near middle (Figure 14.1) 4
 • Proboscis without a white ring near middle 6

4. Abdomen with yellowish-white dorsal median longitudinal
stripe or row of disconnected spots; wing scales entirely dark
or intermixed brown and white..5
 • Abdomen lacking pale median longitudinal stripe, but
possessing transverse basal bands of white scales; wing scales
entirely dark (Figure 14.1) ....*taeniorhynchus*

5. Lateral and dorsal pale markings of abdomen yellowish; first
segment of hind tarsus may or may not have a median ring,
which, if present, is white ....*nigromaculis* (in part)
 • Lateral pale markings of abdomen white, dorsal markings
yellowish; first segment of hind tarsus with a median yellow
ring in addition to basal ring (salt marsh mosquito) ....*sollicitans*

6. Mesonotum lacking a silvery, lyre-shaped marking of scales 7
 • Mesonotum with conspicuous silvery lyre-shaped marking of
scales on a dark background (Figure 14.2) (urban yellow fever
mosquito) ....*aegypti*

7. Basal white rings of tarsal segments narrow...............................8
 • Basal white rings of tarsal segments broad, especially on rear
legs 9

8. Seventh abdominal tergite entirely pale-scaled; lower
mesanepimeral bristles present (see Figure 12.1) (brown salt
marsh mosquito) ....*cantator*
 • Seventh abdominal tergite mostly dark-scaled; lower
mesanepimeral bristles absent ....*vexans*

9. Palpus with some pale scales 10
   • Palpus all dark . . . .*nigromaculis* (in part)

10. Abdominal tergites dark-scaled, with some pale scales basally,
    forming basal bands in some species 11
    • Abdominal tergites yellow-scaled, without basal bands . . . .*flavescens*

11. Pale wing scales scattered generally over wing; torus (enlarged
    basal segment of antennae) with a conspicuous inner patch of
    pale scales 12
    • Pale wing scales confined almost entirely to front part of wing;
    torus with a few inconspicuous pale scales on inner surface . . . .*increpitus*

12. Hind wing margin at base of fringe bordered with mixed black
    and white scales; wing scales mostly very broad (California salt
    marsh mosquito) . . . .*squamiger*
    • Hind wing margin at base of fringe bordered with minute
    unicolored scales; wing scales narrow to moderately broad
    (Figure 14.4) . . . .*fitchii*

13. Wing with dark and white scales intermixed 14
    • Wing with scales all dark or with some white scales on anterior
    veins 15

14. In female, the anal vein with more pale scales than dark ones . . . .*dorsalis*
    • In female, the anal vein with more dark scales than pale
    ones . . . .*melanimon*

15. Palpi with white bands; scutellum with broad white scales
    (western tree hole mosquito) . . . .*sierrensis*
    (formerly known as *varipalpus*)
    • Palpi entirely dark; scutellum with narrow yellowish scales . . . .*atropalpus*

16. Scutum (mesonotum) uniformly dark, without contrasting
    lines or stripes of paler scales 17
    • Scutum with contrasting lines, stripes, or areas of paler (white,
    silver, golden, or reddish) scales 18

17. Abdominal tergites with indistinct or narrow dorsal, basal
    bands of white scales that merge laterally to form a continuous
    white line of scales; procoxa with dark, flat scales (a widely
    distributed cinnamon-brown mosquito of U.S.) . . . .*cinereus*
    • Abdominal tergites with distinct dorsal, basal bands of white
    scales not merging to form a lateral white line; procoxa with
    pale scales . . . .*abserratus*

18. Scutum with one or two median stripes or lines of paler
    scales 19
    • Scutum with lateral areas of contrasting paler scales 22

19. Scutum marked with stripes of white scales 20
    • Scutum marked with two stripes of reddish-brown scales that
    may merge posteriorly (Figure 14.3); dorsal abdominal
    segments with white basal bands (a widespread woodland
    species in U.S.) . . . .*sticticus*

20. Scutum with one broad median stripe of white scales 21
    • Scutum with two broad medial stripes of white scales that
    converge posteriorly (Figure 14.5) . . . .*trivittatus*

21. Occiput of head completely white-scaled (Figure 14.6); white
    stripe of scutum broad; small species (wing length circa 2.5
    mm) ....*dupreei*
 • Occiput of head with medial white stripe of scales bordered by
    areas of dark scales; white stripe of scutum narrow (Figure
    14.7) (0.3 or less mesonotal width); medium-size species (wing
    length circa 3.0 mm) ....*atlanticus*

22. Sides and prescutellar area of thorax lined with golden scales;
    abdomen without basal bands of white scales, or with narrow
    basal bands on only a few segments (a common bog mosquito
    of northeastern U.S.) ....*aurifer*
 • Sides of thorax white, with broad flat, overlapping scales of
    brilliant white or silver; dorsum of abdomen with dark scales
    only (a widespread tree hole species in all except far western
    U.S.) ....*triseriatus*

Key to species of *Culiseta* of America north of Mexico, adults

1. Hind tarsi with white rings on some segments, though
   sometimes very narrow 2
 • Hind tarsi entirely dark 5

2. Wings spotted with dense patches of dark scales; tarsi with
   faint whitish rings at basal end of segments only 3
 • Wings not spotted; tarsi with faint rings at both ends of
   segments ....*morsitans*

3. Scales present on cross veins; hind tarsal ring on segment 2
   covering 0.25 to 0.33 of the segment 4
 • Scales not present on cross veins; hind tarsal ring on segment 2
   covering about 0.1 of the segment ....*incidens*

4. Femora each with subapical white ring ....*particeps*
 • Femora without such rings ....*alaskensis*

5. Costa of wing entirely dark-scaled......................................6
 • Costa of wing with pale scales near base ....*inornata*

6. Cross veins arising from medius separated by more than
   length of either cross vein (see Figure 12.7); spiracular bristles
   dark ....*melanura*
 • Cross veins arising from medius separated by less than length
   of either cross vein; spiracular bristles yellow.........................*impatiens*

Key to commoner *Culex* species of North America, adults

1. Proboscis ringed by light bands 2
 • Proboscis not so ringed 3

2. A fine white line or row of spots on outside of femora and
   tibiae; v-shaped dark markings on venter of each abdominal
   segment ....*tarsalis*
 • Without such a line or row of white spots; black median spot on
   venter of each abdominal segment ....*peus*

3. Tarsi with rather distinct white rings 4
   • Tarsi without distinct white rings, and if narrow rings present,
   they are brownish 6

4. Tarsal rings narrow 5
   • Tarsal rings broad . . . .*coronator*

5. Dark triangular markings on venter of abdominal segments;
   tarsal rings whitish . . . .*thriambus*
   • Dark markings on venter narrow, transverse, at apex of each
   segment . . . .*restuans* (California form)

6. Scales of vein R_{2-3} narrow; occiput usually lacking broad
   appressed scales dorsally 7
   • Scales of vein R_{2-3} broadened; occiput usually with broad
   appressed scales dorsally . . . . .subgenus *Melanoconion*
   (*anips* from southern and Baja California, *abominator*, *atratus*,
   erraticus, *iolambdis*, *mulrennani*, *peccator*, and *pilosus* from
   southeastern U.S. all separated with great difficulty)

7. Abdominal segments with pale scales basal 8
   • Abdominal segments with posterior transverse bands of pale
   scales . . . .*apicalis*

8. Mesonotum not bright reddish-brown 9
   • Mesonotum bright reddish-brown . . . .*erythrothorax*

9. Abdominal segments each with a rather broad basal band of
   whitish scales dorsally 10
   • Abdominal segments usually with narrow, dull yellowish, basal
   bands dorsally, and with apices of the segments blended with
   dull yellowish scales . . . .*salinarius*

10. Mesonotum without whitish dots in front of wing roots 11
    • Mesonotum with 2 whitish dots just in front of wing roots . . . .*restuans*

11. Bands of abdominal tergites not connected with lateral spots;
    mesonotal integument commonly dark brown . . . .*pipiens quinquefasciatus*
    • Bands of abdominal tergites III and IV connected with lateral
    spots; mesonotal integument commonly reddish-brown . . . .*pipiens pipiens*
    pipiens molestus

Key to commoner species of *Psorophora* of North America, adults[a]

1. Wing scales mixed dark and white; hind femur with a more or
   less distinct narrow subapical ring of white scales 2
   • Wing scales all dark or with only a few inconspicuous white
   scales on costa and subcosta; hind femur without a subapical
   ring of white scales 3

2. Segment 1 of hind tarsus with white-scaled rings at base and at
   middle; wings speckled with brown and white scales in no
   definite pattern . . . .*columbiae*
   • Segment 1 of hind tarsus largely pale-scaled; wings with
   definite areas of dark and white scales . . . .*discolor*

3. Medium-size mosquitoes; legs III not very shaggy; femora III
   without long, erect, apical scales; fifth segment of tarsi III
   entirely white 4

- Very large mosquitoes; legs III very shaggy; femora III with long, erect, apical scales; fifth segment of tarsi III not entirely white .... *ciliata*

4. Tarsi of legs III wholly dark-scaled; apical submedian triangular patches of golden-yellow scales on abdominal tergites .... *cyanescens*
- Tarsi of legs III with white on apical segments; apicolateral triangular patches of golden-yellow to whitish-yellow scales on all but first abdominal tergites .... *ferox*

*a*Adapted from Carpenter and LaCasse (1955).

References

Carpenter, S. J., and W. J. LaCasse. 1955. *Mosquitoes of North America* (*North of Mexico*). University of California Press, Berkeley. 360 pp.

Siverly, R. E. 1972. *Mosquitoes of Indiana*. Indiana State Board of Health, Indianapolis. 120 pp.

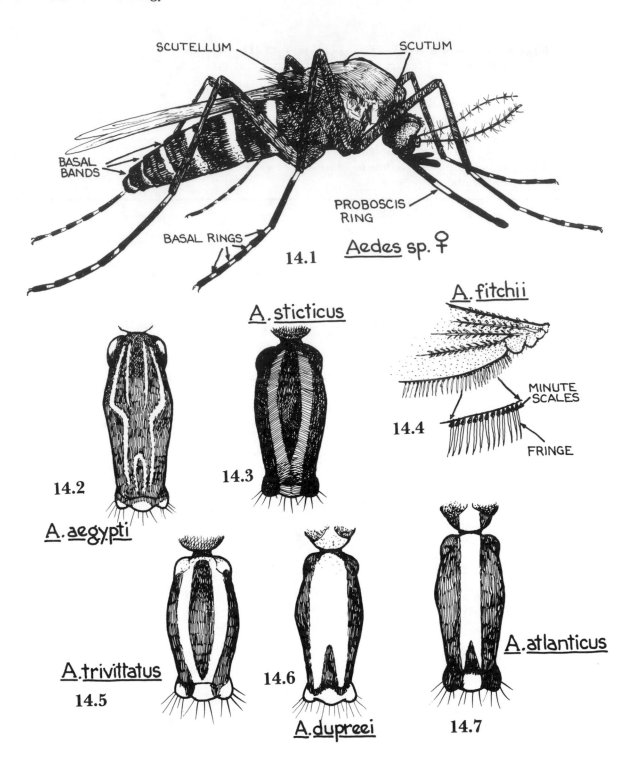

SCUTELLUM

SCUTUM

BASAL BANDS

PROBOSCIS RING

BASAL RINGS

14.1   Aedes sp. ♀

A. sticticus

14.3

A. fitchii

MINUTE SCALES

FRINGE

14.4

14.2

A. aegypti

A. trivittatus

14.5

A. dupreei

14.6

A. atlanticus

14.7

15   Order Diptera: Culicinae (larvae)

Larvae of mosquitoes are studied best as microscope slide mounts under the compound microscope. In preparing culicine larvae for mounting, the abdomen is cut partially through on the dorsal side between the sixth and seventh segments to permit rotation of the posterior section bearing the siphon and anal segment so that it lies with a lateral side uppermost.

Key to commoner species of *Psorophora* of North America, fourth-stage larvae[a]

1. Pecten teeth few (less than 10), not prolonged into hairlike
   filaments; siphonal tuft multiple, large, small or obsolete 2
   - Pecten teeth numerous (about 18 or more), each terminating
     in a hairlike filament; siphonal tuft represented by a single
     long hair . . . .*ciliata*

2. Siphonal tuft small or obsolete, multiple; siphon large, more or
   less inflated medially; antennae not inflated 3
   - Siphonal tuft large, multiple, as long as the siphon; siphon
     small, not inflated; antennae inflated . . . .*discolor*

3. Upper frontal head hair 5 single or double (rarely triple),
   lower frontal 6 single, double, or triple 4
   - Upper frontal and lower frontal head hairs 5 and 6 multiple . . . .*columbiae*

4. Upper frontal and lower frontal hairs 5 and 6 with one or
   more hairs single . . . .*cyanescens*
   - Upper frontal head hair 5 double, lower frontal 6 double or
     triple . . . .*ferox*

[a]Adapted from Carpenter and LaCasse (1955).

Key to commoner species of *Aedes* of North America, fourth-stage larvae

1. Anal segment ringed by saddle 2
   - Anal segment not entirely ringed by the saddle (Figure
     12.28) 4

2. Pecten of air tube without detached teeth outwardly 3
   - Pecten of air tube with detached teeth outwardly (see Figure
     12.30) . . . .*nigromaculis*

3. Lateral abdominal hairs double on segments 3–5; comb scales
   thorn-shaped at tip with subapical spinules not more than
   two-thirds as long as the median spine (Figure 12.30) . . . .*sollicitans*
   - Lateral abdominal hairs triple on segments 3–5; comb scales
     rounded apically and fringed with subequal spinules . . . .*taeniorhynchus*

4. Pecten of siphon with detached teeth outwardly (Figure
   15.2) 5
   - Pecten of siphon with all teeth rather evenly spaced 7

5. Siphon with a hair tuft beyond the pecten (see Figure 12.30) 6
   - Siphon with a hair tuft within the pecten . . . .*atropalpus*

6. Comb scales of 8th segment arranged in 3 or more irregular rows ....*flavescens* (in part)
   - Comb scales of 8th segment arranged in a single or irregular double row (see Figure 12.28) ....*vexans*

7. Antenna smooth; antennal tuft represented by a single hair 8
   - Antenna spiculate; antennal tuft composed of double or multiple hairs (Figure 15.3) 9

8. Eight to 12 comb scales, arranged in a single row (urban yellow fever mosquito) (Figure 12.28) ....*aegypti*
   - Comb scales more numerous, arranged in a triangular patch (western tree hole mosquito) (until recently known as *varipalpus*) ....*sierrensis*

9. Each comb scale with strong median spine and weak subapical spinules not over half as long 10
   - Each comb scale rounded apically and fringed with subequal spinules 11

10. Siphon stout, length usually less than 4 times the width; lateral hair on anal segment shorter than the saddle ....*flavescens* (in part)
    - Siphon slender, length 4 to 5 times the width; lateral hair on anal segment longer than the saddle ....*fitchii*

11. Upper head hair single to triple; lower head hair single or double 12
    - Upper head hair usually with 4 or more branches; lower head hair with 3 or more branches ....*cantator*

12. Lateral hair of anal segment shorter than the saddle 13
    - Lateral hair of anal segment longer than the saddle ....*squamiger*

13. Siphonal tuft located beyond middle of siphon; anal gills short and budlike (sometimes as long as or longer than saddle) 14
    - Siphonal tuft usually located before middle of siphon; anal gills as long as or longer than the saddle ....*increpitus*

14. Mesothoracic hair 1 large; prothoracic hair 1 usually double or triple ....*dorsalis*
    - Mesothoracic hair 1 small; prothoracic hair 1 usually single (see Figure 12.28) ....*melanimon*

Key to species of *Culiseta* of America north of Mexico, fourth-stage larvae

1. Siphon without a row of median tufts along side beyond pecten (Figure 12.39) 2
   - Siphon with such a row of tufts (Figure 15.1) ....*melanura*

2. Pecten of siphon followed by a row of long hairs on outer half 3
   - Pecten of siphon not thus followed by hairs ....*morsitans*

3. Upper frontal and lower frontal head hairs 5 and 6 similar in size and number of branches (see Figure 12.29) ....*impatiens*
   - Lower frontal head hair with fewer branches than hair 5 and usually somewhat longer 4

4. Saddle of anal segment smooth, without coarse spicules at apex 5

- Saddle of anal segment with a prominent group of coarse spicules . . . .*particeps*

5. Lateral hair of anal segment fine, shorter than the saddle 6
- Lateral hair of anal segment somewhat coarse, as long as or longer than anal saddle . . . .*inornata*

6. Mesothoracic hair 1 usually single . . . .*incidens*
- Mesothoracic hair 1 usually multiple (see Figure 12.28) . . . .*alaskensis*

Key to commoner species of *Culex* of North America, fourth-stage larvae

1. Antennae not evenly tapered; antennal tuft inserted near outer third of shaft 2
- Antennae evenly tapered with antennal tuft inserted near middle of shaft . . . .*restuans*

2. Siphon (air tube) with several pairs of multiple hair tufts, usually regularly placed 3
- Siphon with several single hairs, irregularly placed, and at most a single pair of 2 or 3 haired subapical tufts . . . .*thriambus*

3. Lower head hair long, single or double 4
- Lower and upper head hairs long, with 3 or more branches 7

4. Siphon with 2 or 3 pairs of small subdorsal tufts in addition to siphonal tufts . . . . .subgenus *Melanoconion* 5
- Siphon lacking small subdorsal tufts . . . .*apicalis*

5. Comb scales of 8th segment long, thorn-shaped, placed in an irregular single or double row 6
- Comb scales rounded apically and fringed with subequal spinules, arranged in a triangular patch . . . .*peccator*

6. Siphon with 5 or 6 pairs of siphonal tufts, none more than half the siphon length . . . .*erraticus*
- Siphon with about 8 pairs of siphonal tufts, the proximal pair about as long as the siphon . . . .*pilosus*

7. Siphonal tufts not all in a straight line 8
- Siphonal tufts all in a straight line . . . .*tarsalis*

8. Siphon 4 or 5 times as long as wide; lower head hair usually with more than 4 branches 9
- Siphon 6 to 8 times as long as wide; lower head hair usually with 3 or 4 branches 10

9. Lateral hairs of abdominal segments III and IV usually triple; dorsal microsetae toward apex of anal saddle notably larger than those of middorsal area . . . .*peus*
- Lateral hairs of abdominal segments III and IV usually double; dorsal microsetae toward apex of saddle not appreciably larger than those of middorsal area . . . .*pipiens pipiens*
  pipiens molestus
  pipiens quinquefasciatus

10. Anal gills commonly a little longer than anal saddle; normally 4 pairs of siphonal tufts . . . .*salinarius*
- Anal gills about as long as anal saddle; normally 5 pairs of siphonal tufts . . . .*erythrothorax*

Reference

Carpenter, S. J., and W. J. LaCasse. 1955. *Mosquitoes of North America (North of Mexico)*. University of California Press, Berkeley. 360 pp.

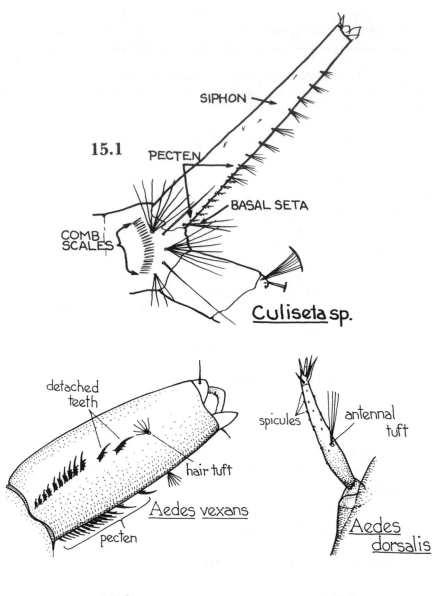

15.1

SIPHON

PECTEN

BASAL SETA

COMB SCALES

Culiseta sp.

detached teeth

hair tuft

pecten

Aedes vexans

15.2

spicules

antennal tuft

Aedes dorsalis

15.3

16 Order Diptera: Tabanidae (horse flies and deer flies)

Tabanidae is one of the largest of dipteran families, with an estimated 8,000 species and worldwide distribution. Tabanids are robust, medium- to large-size flies (body length ranging from 5 to 30 mm). The eyes are large, covering most of the head, and, in most groups, males are holoptic (Figure 16.5). In life many species display colorful eye banding that all but disappears on dried specimens (Figure 16.6). Traces of eye patterns can be revealed by hydrating pinned flies.

Tabanids are strong fliers, and taxonomically the family has clearly defined limits based on typical wing venation (Figure 16.1, 16.2). Females of most species take blood, and both sexes feed on plant juices or excreta. Blood-sucking tabanids are important pests of vertebrate hosts, both by the effects of their direct attacks and by acting as cyclic and mechanical vectors of disease agents. Their importance as vectors has been reviewed by Zumpt (1949).

Mackerras (1955) divides the world's Tabanidae into four subfamilies based primarily on genital characteristics.

Key to subfamilies of American Tabanidae[a]

1. Proboscis and mouthparts minute; palpi small, globular in both sexes (Figure 16.7); 3rd antennal segment with 6–8 annuli (species restricted to S. America and Africa) ....Scepsidinae
   - Proboscis and mouthparts well developed; palpi never small and globular in both sexes 2

2. Third antennal segment with 6–8 distinct annuli (Figures 16.10, 16.17); hind tibial spurs nearly always present; 9th tergite entire in both sexes and style of male hypopygium bifid (Figure 16.12) (species moderately distributed: S. America, Mexico, southern U.S., Mediterranean Coast, eastern Africa, India, and Australia) ....Pangoniinae 4
   - Third antennal segment with never more than 5 distinct annuli (Figure 16.3); 9th tergite divided in both sexes (Figures 16.11, 16.13); style of male hypopygium not bifid 3

3. Ocelli usually absent, small, or rudimentary if present; hind tibial spurs absent; style of male hypopygium truncate (Figure 16.13); caudal ends of the spermathecal ducts of female with mushroomlike expansions (Figure 16.15) (species widely distributed in all temperate and tropical areas of the world) ....Tabaninae 12
   - Ocelli well developed; hind tibial spurs usually well developed (Figure 16.8), sometimes small; style of male hypopygium more or less pointed in dorsal view (Figure 16.11); caudal ends of spermathecal ducts of female never with mushroomlike expansions (Figure 16.14) (species widely distributed in all temperate and tropical areas of the world) ....Chrysopinae 9

Genera of Tabanidae of America north of Mexico, females only
Subfamily Pangoniinae,[b] (Genus *Asaphomyia*, 1 sp. in Texas, not included)

4. Eyes with upper inner angles acute; anterior portion of wings infuscated; frons broader than width of the eye (Figure 16.16) ....*Goniops* (1 sp.) (*G. chrysocoma*)

- Eye angles normal; wings of uniform color 5

5. Apical palpal segment with a pit or groove (Figure 16.19) 6
- Apical palpal segment without a pit or groove 8

6. Proboscis long, subequal to or longer than height of head;
  stump vein present (Figure 16.19) (central, western, and
  southwestern U.S.) ....*Esenbeckia* (3 spp.)
- Proboscis short, length much less than height of head.....................................7

7. Eyes bare or only slightly pilose; apical palpal segment grooved
  (Figure 16.17) (western U.S.) ....*Apatolestes* (11 spp.)
- Eyes and body densely pilose; apical palpal segment with a pit
  (only far western U.S.) ....*Brennania* (2 spp.)

8. Stump vein absent; not known to be hematophagous ....*Stonemyia* (2 spp.)
- Stump vein present (see Figure 16.18) (western U.S.) ....*Pilimas* (3 spp.)

Subfamily Chrysopinae[c]

9. Third antennal segment (flagellum) with 3 divisions
  (2 annuli) ....*Mercomyia* (3 spp.)
- Third antennal segment with 5 divisions (4 annuli) (Figure
  16.9) 10

10. Wings hyaline; scape of antenna at least twice as long as
   pedicle; flagellum longer than scape and pedicle combined
   (Figure 16.20); eyes (in live or relaxed specimens) with small,
   scattered dark spots ....*Silvius* (9 spp.)
- Wings infuscated or with a dark crossband (Figure 16.1); scape
   and pedicle of antenna subequal, or scape only 1.5 times the
   length of pedicle; eyes (in live or relaxed specimens) with large
   angular dark spots (Figure 16.1) 11

11. Antennae slender and elongate, scape 1.5 times the length of
   the pedicle (Figure 16.21); abdomen globose, much wider than
   thorax; wings evenly infuscated ....*Neochrysops* (1 spp.) (*N. globosus*)
- Antennae variable (slender or swollen), pedicle and scape
   usually subequal (Figure 16.9); abdomen not globose, subequal
   to width of thorax; wings irregularly infuscated (or "pictured"
   – may be nearly hyaline in a few species) ....*Chrysops* (75 spp.)

Subfamily Tabaninae[c]

12. Frons without median callus or without basal callus, or without
   both median and basal callus 13
- Frons with median and basal callus (Figure 16.22) (may be
   fused: Figure 16.3) 14

13. Eye pilose, with diagonal purplish "band"...........................*Atylotus* (7 spp.)
- Eye bare; body compact, uniformly greenish-yellow ....*Chlorotabanus* (1 sp.)
   (*C. crepuscularis*)

14. Dorsal base of 3rd antennal segment with forward-projecting,
   hooklike process (Figure 16.4, 16.23); eye sparsely pilose 15
- Third antennal segment with or without such a process, but if
   process present, eye is bare (i.e., eye may be pilose without the
   process, or bare with or without the process) 16

15. Subcallus (see Figure 16.22) with lateral erect black hairs (Gulf
   Coast states) ....*Agkistrocerus* (2 spp.)

- Subcallus without such hairs (Mississippi River Valley and Gulf Coast states) .... *Hamatabanus* (3 spp.)

16. Scape stout and long (but more than 2 times as long as wide) (Figure 16.24); dark wings marked with irregular hyaline bands and spots .... *Haematopota* (5 spp.)
    - Scape length and width subequal; wings hyaline, infuscated, or variously marked 17

17. Vertex with distinct ocellar tubercle (Figure 16.3) 18
    - Vertex without ocellar tubercle 19

18. Eye bare; frons narrow; basal callus a swelling at base of a slender ridge (median callus) (southeastern U.S. and Arizona) .... *Leucotabanus* (2 spp.)
    - Eye pilose; frons of variable width, basal callus and median callus usually broad (Figure 16.3) (with many widely distributed species) .... *Hybomitra* (54 spp.)

19. Subcallus large (swollen) and shiny; genae denuded and shiny .... *Bolbodimyia* (1 sp.), *B. atrata*
    Whitneyomyia (1 sp.), *W. beautifica*
    (see Stone, 1938, for separation)
    - Subcallus not enlarged, or if so, then genae are pollinose 20

20. Base of 3rd antennal segment without a dorsal angle; frons very narrow, median callus a slender line; wing with at least a subapical brown spot (southeastern coastal states, "the yellow fly of the dismal swamp") .... *Diachlorus* (1 sp.) (*D. ferrugatus*)
    - Without the above combination of characters 21

21. Flagellum (3rd antennal segment) of antenna with 3 divisions (2 annuli); small fly, 7–9 mm long (eastern U.S. coastal states) .... *Microtabanus* (1 sp.)
    (*M. pygmaeus*)
    - Flagellum of antenna usually with 5 divisions (4 annuli) 22

22. Antennae and palpi with erect hairs; palpus short and blunt; proboscis small .... *Anacimus* (3 spp.)
    - Without the above combination of characters 23

23. Basal division of third antennal segment usually rounded above (Figure 16.26) (if there is suggestion of a dorsal angle, frons is widened below; Figure 16.25); small, slender flies (10–14 mm long); stump vein usually present at base of R$_4$ (representatives across southern U.S.) .... *Stenotabanus* (7 spp.)
    - Basal division of 3rd antennal segment with distinct dorsal angle (usually excised); frons never widened below; small to large flies (9–30 mm); stump vein may or may not be present .... *Tabanus* (94 spp.)

[a]Modified from Mackerras (1955).
[b]Modified from Brennan (1935) and Middlekauff and Lane (1980).
[c]Modified from Jones and Anthony (1964) and Middlekauff and Lane (1980).

References

Brennan, J. M. 1935. The Pangoniinae of Nearctic America (Tabanidae, Diptera). *Univ. Kansas Sci. Bull. 22*:249–401.

Frost, S. W., and L. L. Pechuman. 1958. The Tabanidae of Pennsylvania. *Trans. Am. Entomol. Soc. 84*:169–217.

Hays, K. L. 1956. *A Synopsis of the Tabanidae (Diptera) of Michigan.* University of Michigan Museum of Zoology Misc. Publ. 98. 79 pp.

Jones, C. M., and D. W. Anthony. 1964. *The Tabanidae (Diptera) of Florida.* U.S. Dept. of Agriculture Tech. Bull. 1295. 85 pp.

Mackerras, I. M. 1954. The classification and distribution of Tabanidae (Diptera). I. General review. *Australian J. Zool.* 2:431–54.

– 1955. The classification and distribution of Tabanidae (Diptera). II. History: morphology: classification: subfamily Pangoniinae. *Australian J. Zool.* 3:439–511.

Middlekauff, W. W., and R. S. Lane. 1980. *Adult and Immature Tabanidae (Diptera) of California.* Bull. California Insect Survey, 22. University of California Press, Berkeley. 99 pp.

Pechuman, L. L., 1957. The Tabanidae of New York. *Rochester Acad. Sci. 10*:121–79.

Pechuman, L. L., H. J. Tesky, and D. M. Davies. 1961. The Tabanidae (Diptera) of Ontario. *Proc. Entomol. Soc. Ontario* (1960) *91*:77–121.

Philip, C. B. 1931. *The Tabanidae (Horseflies) of Minnesota.* University of Minnesota Agr. Exp. Sta. Tech. Bull. 80. 132 pp.

– 1955. New North American Tabanidae. IX. Notes on and Keys to the genus *Chrysops* Meigen. *Rev. Brazil. Entomol. 3*:47–128.

Richards, L. L., and K. L. Knight. 1967. The horse flies and deer flies of Iowa (Diptera: Tabanidae). *Iowa State J. Sci. 41*:313–62.

Roberts, R. H., and R. J. Dicke. 1958. Wisconsin Tabanidae. *Trans. Wisc. Acad. Sci., Arts and Letters 47*:23–42.

Stone, A. 1938. *The Horseflies of the Subfamily Tabanidae of the Nearctic Region.* U.S. Dept. of Agriculture Misc. Publ. 305. 172 pp.

Thompson, P. H. 1967. Tabanidae of Maryland. *Trans. Am. Entomol. Soc. 93*:463–519.

Zumpt, F. 1949. Medical and veterinary importance of horseflies. *So. African Med. J. 23*:359–62.

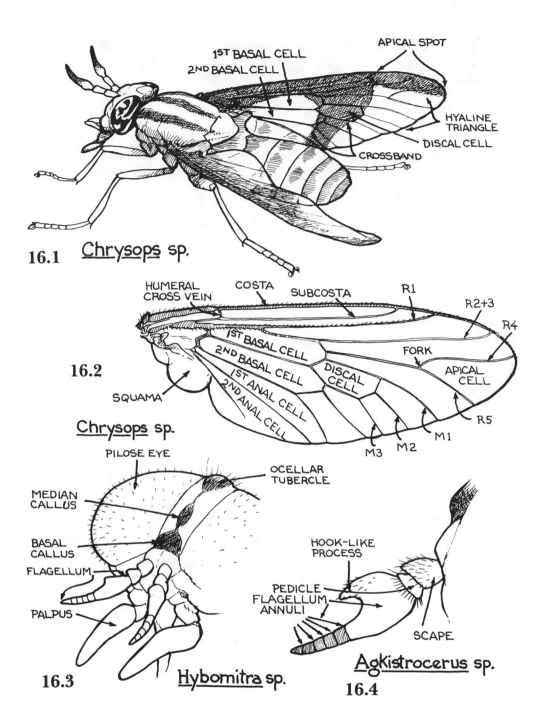

16.1 _Chrysops_ sp.

16.2 _Chrysops_ sp.

16.3 _Hybomitra_ sp.

16.4 _Agkistrocerus_ sp.

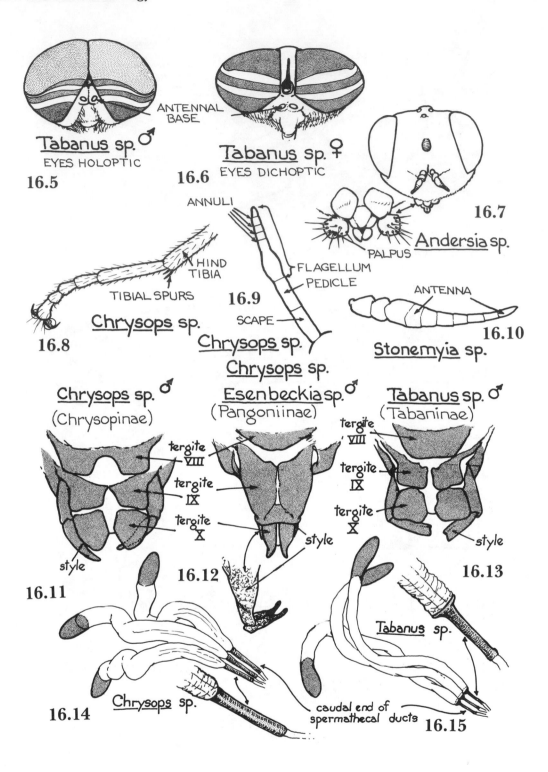

Tabanus sp. ♂
EYES HOLOPTIC
16.5

ANTENNAL BASE

Tabanus sp. ♀
EYES DICHOPTIC
16.6

16.7
Andersia sp.

PALPUS

ANNULI

HIND TIBIA

TIBIAL SPURS
16.8
Chrysops sp.

FLAGELLUM
PEDICLE
16.9
Chrysops sp.
SCAPE
Chrysops sp.

ANTENNA
16.10
Stonemyia sp.

Chrysops sp. ♂
(Chrysopinae)

Esenbeckia sp. ♂
(Pangoniinae)

Tabanus sp. ♂
(Tabaninae)

tergite VIII
tergite IX
tergite X
style
16.11

style
16.12

tergite VIII
tergite IX
tergite X
style
16.13

16.14
Chrysops sp.

Tabanus sp.

caudal end of spermathecal ducts
16.15

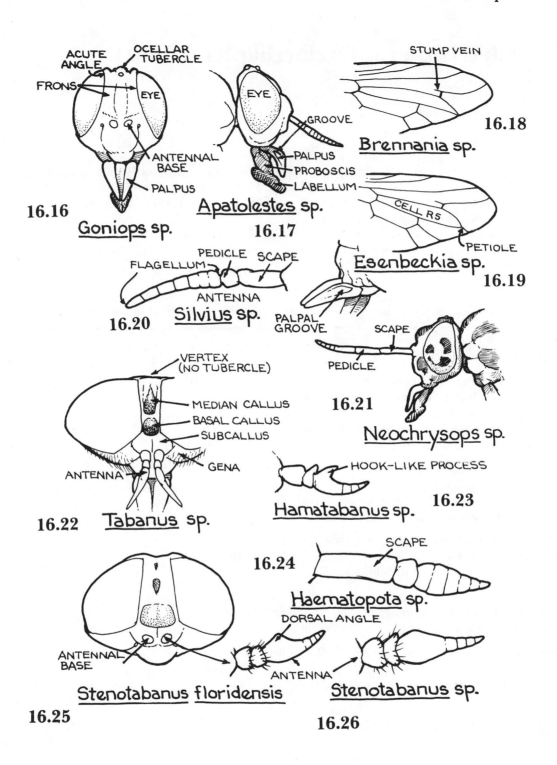

ACUTE ANGLE
OCELLAR TUBERCLE
FRONS
EYE
ANTENNAL BASE
PALPUS
16.16
Goniops sp.

EYE
GROOVE
PALPUS
PROBOSCIS
LABELLUM
Apatolestes sp.
16.17

STUMP VEIN
16.18
Brennania sp.

CELL R5
PETIOLE
Esenbeckia sp.
16.19

FLAGELLUM
PEDICLE
SCAPE
ANTENNA
16.20 **Silvius sp.**

PALPAL GROOVE

SCAPE
PEDICLE
16.21
Neochrysops sp.

VERTEX (NO TUBERCLE)
MEDIAN CALLUS
BASAL CALLUS
SUBCALLUS
GENA
ANTENNA
16.22 **Tabanus sp.**

HOOK-LIKE PROCESS
16.23
Hamatabanus sp.

16.24
SCAPE
Haematopota sp.

ANTENNAL BASE
DORSAL ANGLE
ANTENNA
Stenotabanus floridensis
16.25

Stenotabanus sp.
16.26

17   Order Diptera: Cyclorrhapha (adults)

The Cyclorrhapha includes some 65 families and constitutes those Diptera known as the higher flies. In contrast to the Nematocera and Brachycera, few higher flies are obligate blood feeders as adults, but many are associated with organic waste in the form of carrion and fecal matter. Many of these are protelean parasites, among them the maggots causing myiasis.

The 20 or so families characterized in the following keys include not only those of medical importance, but also groups that because of superficial similarities or association might be confused with the medically important forms.

Key to certain families, genera, and species of Cyclorrhapha of medical importance, adults

1. Wings present, rarely reduced 2
   - Wings absent or greatly reduced 22

2. Vena spuria absent; anal cell usually shorter than cell M
   (2nd basal) or absent; if anal cell elongate, then a ptilinal suture
   present (Figure 17.6) 3
   - Vena spuria present between R and M (Figure 17.1); anal cell
   longer than cell M; no ptilinal suture . . . . .Syrphidae

3. Hind coxae approximated at base, the legs thus ventrally
   attached 4
   - Hind coxae well separated by a broad sternal plate, the legs
   thus more laterally attached 22

4. Oral openings and mouthparts very small or appearing
   vestigial in undissected specimens (Figure 17.7) (the botflies) 5
   - Mouthparts obviously well developed 9

5. Arista bare; mouthparts appearing vestigial (Figure 17.2) 6
   - Arista plumose; mouthparts apparently functional but small,
   often hidden in beard; very large, robust flies . . . . .Cuterebridae 48

6. Apical cell greatly narrowed (Figure 17.2), or closed at its
   apex 7
   - Apical cell gradually broadening to wing margin . . . . .Gasterophilidae
   (horse stomach bots) *Gasterophilus* 49

7. Apical cell closed and petiolate (Figure 17.3) . . . . .Oestridae (in part)
   (sheep bot fly) *Oestrus ovis*
   - Apical cell open 8

8. Marginally directed spur vein at cubitulus (Figure 17.2)   . . . . .Oestridae (in part)
   (deer nasal bot fly) *Cephenemyia*
   - Cubitulus without spur vein . . . . .Oestridae, Hypodermatinae (heel fly
   of cattle or cattle grub) *Hypoderma* 47

9. Squamae small or vestigial; transverse suture incomplete;
   posterior calli not differentiated; 2nd antennal segment
   without an outer dorsal longitudinal suture 10
   - Squamae well developed (Figure 17.5); transverse suture

complete (Figures 17.4, 17.8) posterior calli well defined by a depression from behind base of wings to above base of scutellum; 2nd antennal segment with an outer dorsal longitudinal suture 16

10. Posterior (metathoracic) spiracle with pubescence only 11
    • Posterior spiracle with at least one bristle on posterior border in addition to pubescence ....Sepsidae

11. Subcosta incomplete, often fused with R_1, or not ending in the costa, or the apex curved forward beyond the bend, often evanescent at the tip 12
    • Subcosta complete, free from the radius, and ending in costa 14

12. Subcosta curved forward at almost a right angle and weakened beyond bend; costa fractured at their junction; wings usually pictured ....Tephritidae (= Trypetidae)
    • Subcosta not so curved or weakened; wings rarely pictured 13

13. Ocellar triangle large; $Cu_1 + M_{3+4}$ vein with a distinct curvature near middle of discal cell (1st M_2) ....Chloropidae (= Oscinidae)
    • Ocellar triangle small; $Cu_1 + M_{3+4}$ straight or not distinctly curved ....Drosophilidae (in part)

14. Oral vibrissae present (Figure 17.6) 15
    • Oral vibrissae absent; wings pictured ....Otitidae (= Ortalidae)

15. Second basal (= 2nd M) and discal (= 1st M_2) cells separated (Figure 17.16) ....Piophilidae
    • Second basal and discal cells confluent (Figure 17.17) ....Drosophilidae (in part)

16. Postscutellum absent or not developed, or if developed then hypopleura with hairs only (Figure 17.10) 17
    • Postscutellum developed as a strong convexity below the scutellum (Figure 17.9); hypopleura with strong bristles ....Tachinidae (= Larvaevoridae)

17. Abdomen tesellated, metallic green, blue, or dull; hypopleura with a row of bristles 20
    • Abdomen usually dull or tan-colored; hypopleura bare or with fine, usually short hairs 18

18. Arista plumose, rays feathered (Figure 17.12); proboscis rigid, porrect; palpi long, ensheathing proboscis at rest; prosternum membranous ....Glossinidae tsetse flies, 1 genus, *Glossina*
    • Arista if plumose with unfeathered rays, proboscis haustellate or, if porrect, not ensheathed by palpi; prosternum not entirely membranous 19

19. Vein 6 long, reaching wing margin, even if only as a fold; usually 1 sternopleural bristle, or undersurface of scutellum with fine, erect hairs (56 genera in N. America) ....Anthomyiidae
    • Vein 6 short, never reaching wing margin; usually more than 1 sternopleural bristle; undersurface of scutellum usually without fine hairs (58 genera in N. America) ....Muscidae 36

20. Two notopleural bristles, occasionally a weak third (Figures 17.8, 17.11); body often metallic 21
    • Three or four notopleural bristles; body not metallic (flesh flies) ....Sarcophagidae (in part) 33

21. Abdomen tesselated or with conspicuous round black spots . . . . Sarcophagidae (in part)33
 - Abdomen metallic (blow flies) . . . . Calliphoridae 25

22. Basal tarsal segment longer than tarsal segments 2 and 3
    combined 23
 - Basal tarsal segment shorter than tarsal segments 2 and 3
    combined 24

23. Basal tarsal segment longer than the remainder of tarsus; head
    small and narrow, folding back into a groove on the
    mesonotum (Figure 17.27); on bats (5 spp. in N. America) . . . . Nycteribiidae
 - Basal tarsal segment subequal in length to terminal tarsal
    segment; head protruding, not folding into mesonotum
    (Figure 17.24); on colonial tree bats (in New Zealand, 1 sp.) . . . . Mystacinobiidae

24. Palpi broader than long; wings uniformly veined; on bats
    (6 spp. in N. America) (Figure 17.25) . . . . Streblidae
 - Palpi elongate, forming a sheath for the proboscis; wing veins
    crowded anteriorly (Figure 17.26); on birds and mammals (12
    genera in N. America) . . . . Hippoboscidae 65

Calliphoridae

25. Stem vein (base of radius before humeral cross vein) ciliate
    posteriorly on upper surface of wing . . . . Chrysomyinae 27
 - Stem vein bare posteriorly on upper surface of wing (Figure
    17.5) 26

26. Prosternum and propleuron bare; ground color dull black or
    at most subshining, never metallic blue, green, or bronze . . . . Polleniinae
 - Prosternum and propleuron setulose or pilose; in the
    California species, at least, the abdomen metallic blue, green,
    or bronze, the sheen sometimes dulled by pollen . . . . Calliphorinae 31

27. Hind coxae pilose posteriorly; mesonotum blue or green, with
    distinct longitudinal black stripes that are clearly visible; lower
    part of head yellow 28
 - Hind coxae bare posteriorly; mesonotum dull, dark blue or
    black, with, at most, stripes that are dependent for visibility on
    proper light incidence; head predominantly dark in color 29

28. Palpus short, filiform; epistoma distinctly narrower than the
    clypeus (see Figure 17.6) . . . . *Cochliomyia* 51
 - Palpus elongate, clavate; epistoma only slightly narrower than
    the clypeus . . . . *Paralucilia wheeleri*

29. Mesonotum convex; anterior (mesothoracic) spiracle with
    bright orange hair . . . . *Phormia*
    (black blowfly) *Phormia regina*
 - Mesonotum and scutellum conspicuously flattened; anterior
    spiracle with dark hair 30

30. Presutural acrosticals (central setae of mesonotum) absent;
    head at vibrissa distinctly longer than at antennae; upper
    squamal lobe black-setose above . . . . *Protophormia*
 - Presutural acrosticals present (see Figure 17.8); head at
    vibrissa no longer than at antenna, upper squamal lobe
    without black hair above (bird nest screw worm flies) . . . . *Protocalliphora* (= *Apaulina*)

31. Lower squamal lobe bare above (see Figure 17.5); abdomen green or bronzy . . . . .tribe Luciliini 52
 • Lower squamal lobe pilose above; abdomen blue green to deep blue 32

32. Scutellum with 3 strong lateral bristles . . . .*Cynomyopsis*
Cynomyopsis cadaverina
 • Scutellum with at least 4 strong lateral bristles . . . .*Calliphora* 56

Sarcophagidae

33. Arista plumose (Figure 17.13); if pubescent or bare, then hind coxa with a fringe of hairs on posterior margin; coxopleural "streak" absent (28 genera in N. America including *Sarcophaga* and *Blaesoxipha,* each with some myiasis-causing species) . . . . .Sarcophaginae
 • Arista bare or pubescent (Figure 17.14), or if plumose, then hind coxa bare on posterior margin; coxopleural "streak" present (Figure 17.14) (20 genera in N. America) . . . . .Miltogramminae 34

34. Meron with a row of bristles and no hairs; tegula usually black, contrasting with pale basicosta (24 genera, 82 spp.) . . . . .Miltogrammini
(typical genus) *Hilarella*
 • Meron with hairs and row of bristles; tegula pale yellow to brown, concolorous with basicosta; palpus yellow (6 genera, 13 spp.) . . . . .Paramacronychiini 35

35. Hind tibia with strong apical posterodorsal bristle; arista bare; third antennal segment dark, brown to black (4 spp. widespread) . . . .*Brachicoma*
 • Hind tibia without or with very weak posterodorsal bristle; third antennal segment orange to yellow (2 spp., causing vertebrate myiasis) . . . .*Wohlfahrtia*

Muscidae

36. Vein 6 reaching less than half the distance to wing margin; vein 7 strongly curved so that if longer, would cross imaginary extension of vein 6 before or at wing margin (Figure 17.18) . . . .(Faniinae) (over 200 spp.),
e.g., *Fannia* 60
 • Vein 6 extending more than half the distance to wing margin; vein 7 weakly curved, so if longer, would not cross an extension of vein 6 37

37. Labella small; proboscis of the piercing type, rigid, porrect, shiny, and nonretractile (Figure 17.4) . . . . .Stomoxyini 38
 • Labella large; proboscis of the sponging type, flexing in retraction 40

38. Maxillary palpi less than half as long as proboscis (Figure 17.4) . . . .*Stomoxys* (e.g., *S. calcitrans,* stable fly)
 • Maxillary palpi subequal in length to proboscis (Figure 17.22) 39

39. Hairs of bucca (cheeks) black; 3rd antennal segment with squared corners . . . .*Lyperosiops*
(e.g., *L. alcis,* moose fly)
 • Hairs of bucca golden; 3rd antennal segment with rounded corners . . . .*Haematobia*
(e.g., *H. Irritans,* horn fly)

40. Mesoanepimeron (pteropleuron) bare 41
- Mesoanepimeron (pteropleuron) setose...43

41. Lower squama enlarged to medially crowd base of scutellum,
caudal margin transverse; vein 4 curved smoothly forward,
ending near or behind apex of wing (Figure 17.19); all tibia
and apical part of mid and hind femora brown to yellow; palpi,
epaulet, and basicosta yellow ....(Muscinae in part) *Muscina* 64
e.g., false stable fly, *M. stabulans*

- Lower squama not enlarged to crowd base of scutelum, caudal
margin semicircular; hind tibia with one or more posterodorsal
bristles ....(Phaoniinae) 42

42. Eyes extend below level of vibrissae, making cheeks very
narrow; profemur of male with distinct concavity in preapical
region ....*Hydrotaea*
- Eyes subequal in height to length of protibia; profemur of
male without a concavity; black, shiny compact
species ....*Ophyra*

43. Propleuron haired; vein 4 bent sharply forward to join margin
near vein 3 at or before apex of wing (Figure 17.15); body
nonmetallic ....(Muscinae in part) *Musca*
e.g., house fly, *M. domestica*

- Propleuron bare; vein 4 curved or bent; proboscis short with
large labella 44

44. Frons bright metallic green...45
- Frons not as above 46

45. Vein 4 sharply bent forward, ending before wing apex;
metallic green species ....*Orthellia*
- Vein 4 weakly curved forward, ending well behind wing apex;
some as nestling bot flies of cavity-dwelling birds ....*Philornis*

46. Midtibia with a strong midposteroventral bristle; vein 4 curved
gently forward, metallic blue-green species ....*Pyrellia*
- Midtibia without such a strong bristle; 4 postsutural bristles;
grayish or metallic spotted species ....*Morellia*

Hypoderma

47. Mesonotum conspicuously reddish-yellow pilose in front, black
behind; scutellum only very indistinctly notched at the apex
(Figure 17.21) (northern cattle grub) ....*H. bovis*
- Mesonotum uniformly yellow-pilose; scutellum slightly, but
distinctly, notched at the apex (Figure 17.20) (common cattle
grub)...*H. lineatum*

Cuterebridae

48. Carina of face distinct; tarsi broad and flattened; sides of
thorax densely pilose; large species, over 18 mm (New World
bot flies on rodents, lagomorphs, and primates, more than 40
spp. in N. America) ....*Cuterebra*
- Face without carina; tarsi not broad or flattened; sides of
thorax not densely pilose; medium size, 15 mm or less (New
World bot fly on a wide range of hosts) ....*Dermatobia*
Torsalo *D. hominis*

Gasterophilus

49. Wings unclouded, clear or suffused dusky but without definite
    bands or spots 50
    - Wings clouded with a transverse brownish band at about the
      middle and 2 brownish spots near apex; abdomen with
      yellowish-brown pile, usually with rows of blackish spots (horse
      bot fly) ....*G. intestinalis* (= *equi*)

50. Vein M_3 distant by much more than its length from cross vein
    r-m; tip of abdomen with orange-reddish pile (nose bot fly) ....*G. haemorrhoidalis*
    - Vein M_3 nearly meeting the cross vein r-m; tip of abdomen
      with yellow pile (Throat botfly) ....*G. nasalis* (= *veterinus*)

Cochliomyia

51. Basicostal scale (see Figure 17.5) black in female and also
    normally in male; lower parafrontal of head with dark hair
    (see Figure 17.6); in both sexes 4th visible abdominal segment
    without well-defined silvery-gray lateral patches when seen
    from above ....(primary screw worm)
    C. hominivorax
    - Basicostal scale whitish to yellow-orange in female; lower
      parafrontal with pale yellowish hair; in both sexes 4th segment
      with well-defined, widely separated patches ....(secondary screw worm)
      C. macellaria

Luciliini

52. Subcostal sclerite (small horny sclerite on lower surface of wing
    at base of subcosta) bare; ocellar triangle not reaching halfway
    to the lunule in the female 53
    - Subcostal sclerite with stiff black setulae; ocellar triangle large,
      reaching about halfway to the lunule in the female ....*Lucilia*

53. Anterior–posterior length of head at antenna and at vibrissa
    more than half head height ....*Bufolucilia*
    - Length of head at antenna and at vibrissa less than half head
      height ....*Phaenicia* 54

54. Basicostal scale yellow to yellowish-white; normally 3
    postsutural acrosticals 55
    - Basicostal scale blackened or dark brown; normally 2
      postsutural acrosticals ....*Phaenicia mexicana*
      (= *Lucilia caesar*)

55. Abdomen varying from bright green to coppery; frontal stripe
    about twice as wide as parafrontal (see Figure 17.6); femora of
    prothoracic legs normally dark metallic blue to black; central
    occipital area with a group of usually about 6 to 8 setae (varies
    from 3 to 14) ....*Phaenicia sericata*
    - Abdomen usually strongly coppery; frontal stripe about as
      wide as parafrontal (sometimes almost twice as wide in
      females); femora of prothoracic legs normally metallic green;
      central occipital area usually with 1 seta (sometimes 2) ....*Phaenicia cuprina* (= *pallecens*)

Calliphora

56. Three postsutural intraalars 57
    - Two postsutural intraalars (Figure 17.8) 58

57. Bucca (cheeks) reddish (Figure 17.6) . . . . *C. coloradensis*
 ● Bucca, when fully horny, black . . . . *C. livida*

58. Basicosta black; bucca (cheeks), when fully colored, black 59
 ● Basicosta usually yellow to orange yellow; bucca reddish on at
   least anterior half . . . . *C. vicina*

59. Bucca with hair mostly or wholly black . . . . *C. terrae-novae*
 ● Bucca with hair mostly reddish-orange or reddish-yellow
   posteriorly . . . . *C. vomitoria*

Fannia

60. Metacoxa with 1 or 2 posteroventral bristles; third antennal
    segment pale 61
 ● Metacoxa lacking posteroventral bristles; third antennal
   segment black . . . . *F. thelaziae*

61. Thorax faintly 3 striped; sides and base of abdomen pale 62
 ● Thorax dark brown, not striped; abdomen gray, dark,
   pollenose basally 63

62. Legs black . . . . (lesser house fly) *F. canicularis*
 ● Legs pale or yellow except for black tarsi . . . . *F. conspicua*

63. Legs black; thorax dark gray; median abdominal spot
    parallel-sided . . . . (latrine fly) *F. scalaris*
 ● Legs pale or yellow except for black tarsi; thorax ash gray,
   pollenose; median abdominal spot subtriangular . . . . *F. benjamini*

Muscina

64. Legs in part reddish-brown . . . . *M. stabulans*
 ● Legs wholly black . . . . *M. assimilis*

Hippoboscidae

65. Wings absent or much reduced in size and functionless 66
 ● Wings of normal shape, used for flight (Figure 17.26) 68

66. Wings reduced to basal stumps; halteres present; on deer from
    western N. America 67
 ● Wings absent; halteres absent (sheep ked) . . . . *Melophagus ovinus*

67. Anterior coxa with a dorsal retrograde spur (Figure 17.23);
    basal sternite of abdomen with convex hind margin . . . . *Neolipoptena ferrisi*
 ● Anterior coxa without dorsal retrograde spur; basal sternite of
   abdomen with concave hind margin . . . . *Lipoptena depressa*

68. Apical tooth of claw simple (claw bidentate); on deer 67
 ● Apical tooth of claw bifid (claw tridentate); on birds 69

69. Wing with 1 or 2 cross veins and an open anal cell 70
 ● Wing with 3 cross veins and a closed anal cell . . . . *Stilbometopa*

70. Wing with only 1 cross vein . . . . *Pseudolynchia*
 ● Wing with 2 cross veins . . . . *Lynchia*

References

Bequaert, J. S. 1953-7. *The Hippoboscidae or Louse-Flies (Diptera) of Mammals and Birds. Entomologica Americana,* vols. 32-37. Entomological Society, Brooklyn, New York, 1053 pp.

Borror, D. J., and D. M. DeLong. 1964. *An Introduction to the Study of Insects,* rev. ed. Holt, Rinehart & Winston, New York.

Chillcott, J. G. 1960. A revision of the nearctic species of Faniinac (Diptera: Muscidae). *Canad. Entomol.,* suppl. 14. 295 pp.

Curran, C. H. 1934. *The Families and Genera of North American Diptera.* American Museum of Natural History, New York.

Downes, W. L., Jr. 1955. Notes on the morphology and classification of the Sarcophagidae and other Calyptrates (Diptera). *Proc. Iowa Acad. Sci.* 62:514-38.

Elderidge, B. F., and M. T. James. 1957. *Typical Muscid Flies of California.* Bull. Calif. Insect Survey 6. 17 pp.

Hall, D. G. 1947. *The Blowflies of North America.* Thomas Say Foundation, Lafayette, Ind.

Huckett, H. C. 1965. *The Muscidae of Northern Canada, Alaska and Greenland (Diptera).* Mem. Entomological Society of Canada, Ottawa, 42. 369 pp.

- 1975. *The Muscidae of California. Exclusive of Subfamilies Muscinae and Stomoxyinae.* Bull. Calif. Insect Survey, 18. 148 pp.

James, M. T. 1947. *The Flies That Cause Myiasis in Man.* U.S. Dept. of Agriculture Misc. Publ. No. 631.

- 1955. *The Blowflies of California (Diptera: Calliphoridae).* Bull. California Insect Survey, vol. 4, no. 1.

Stone, A., C. W. Sabrosky, W. W. Worth, R. H. Foote, and J. R. Coulson. 1965. *A Catalog of the Diptera of America North of Mexico.* U.S. Department of Agriculture, Washington, D.C.

West, L. S. 1951. *The Housefly.* Comstock, New York.

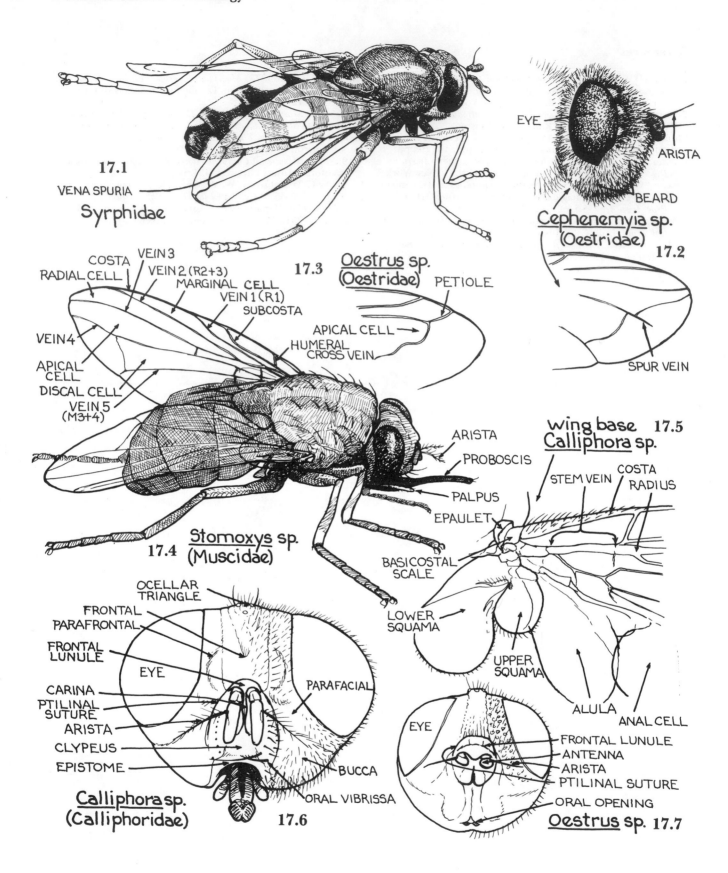

17.1
VENA SPURIA
Syrphidae

EYE
ARISTA
BEARD
Cephenemyia sp.
(Oestridae)
17.2

COSTA
VEIN 3
RADIAL CELL
VEIN 2 (R2+3)
MARGINAL CELL
VEIN 1 (R1)
SUBCOSTA
VEIN 4
APICAL CELL
DISCAL CELL
VEIN 5 (M3+4)
HUMERAL CROSS VEIN

17.3
Oestrus sp.
(Oestridae)
PETIOLE
APICAL CELL
SPUR VEIN

ARISTA
PROBOSCIS
PALPUS
EPAULET
BASICOSTAL SCALE

Stomoxys sp.
(Muscidae)
17.4

wing base 17.5
Calliphora sp.
STEM VEIN COSTA RADIUS
LOWER SQUAMA
UPPER SQUAMA
ALULA
ANAL CELL

OCELLAR TRIANGLE
FRONTAL
PARAFRONTAL
FRONTAL LUNULE
EYE
CARINA
PTILINAL SUTURE
ARISTA
CLYPEUS
EPISTOME
PARAFACIAL
BUCCA
ORAL VIBRISSA
Calliphora sp.
(Calliphoridae)
17.6

EYE
FRONTAL LUNULE
ANTENNA
ARISTA
PTILINAL SUTURE
ORAL OPENING
Oestrus sp. 17.7

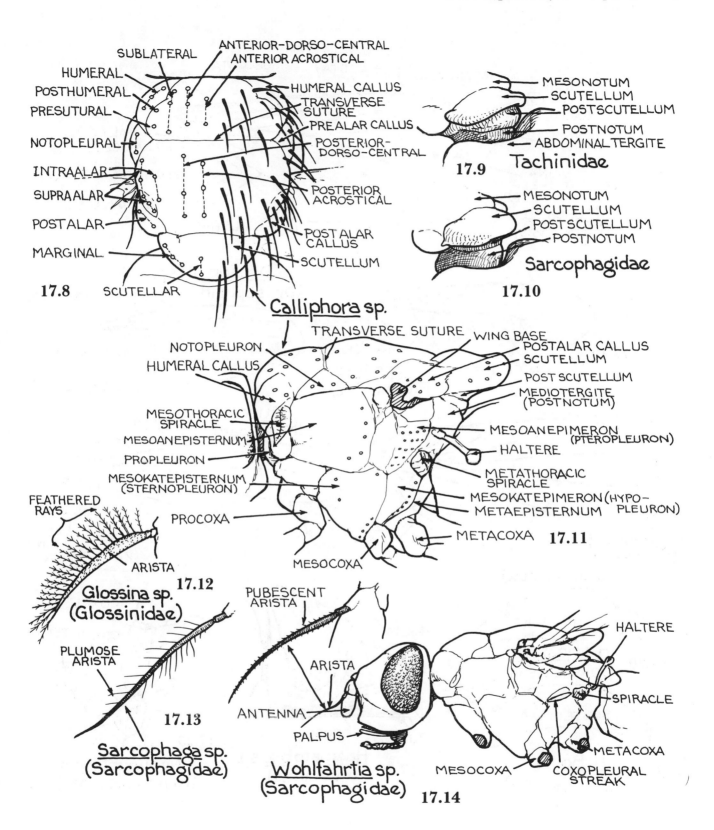

SUBLATERAL
ANTERIOR-DORSO-CENTRAL
ANTERIOR ACROSTICAL
HUMERAL
POSTHUMERAL
PRESUTURAL
NOTOPLEURAL
INTRAALAR
SUPRAALAR
POSTALAR
MARGINAL
SCUTELLAR

HUMERAL CALLUS
TRANSVERSE SUTURE
PREALAR CALLUS
POSTERIOR-DORSO-CENTRAL
POSTERIOR ACROSTICAL
POSTALAR CALLUS
SCUTELLUM

17.8

Calliphora sp.

MESONOTUM
SCUTELLUM
POSTSCUTELLUM
POSTNOTUM
ABDOMINAL TERGITE

17.9 Tachinidae

MESONOTUM
SCUTELLUM
POSTSCUTELLUM
POSTNOTUM

17.10 Sarcophagidae

TRANSVERSE SUTURE
WING BASE
POSTALAR CALLUS
SCUTELLUM
POST SCUTELLUM
MEDIOTERGITE (POSTNOTUM)
MESOANEPIMERON (PTEROPLEURON)
HALTERE
METATHORACIC SPIRACLE
MESOKATEPIMERON (HYPO-METAEPISTERNUM PLEURON)
METACOXA

NOTOPLEURON
HUMERAL CALLUS
MESOTHORACIC SPIRACLE
MESOANEPISTERNUM
PROPLEURON
MESOKATEPISTERNUM (STERNOPLEURON)
PROCOXA
MESOCOXA

17.11

FEATHERED RAYS
ARISTA

Glossina sp. **17.12**
(Glossinidae)

PLUMOSE ARISTA

17.13

Sarcophaga sp.
(Sarcophagidae)

PUBESCENT ARISTA
ARISTA
ANTENNA
PALPUS

HALTERE
SPIRACLE
METACOXA
COXOPLEURAL STREAK
MESOCOXA

Wohlfahrtia sp.
(Sarcophagidae) **17.14**

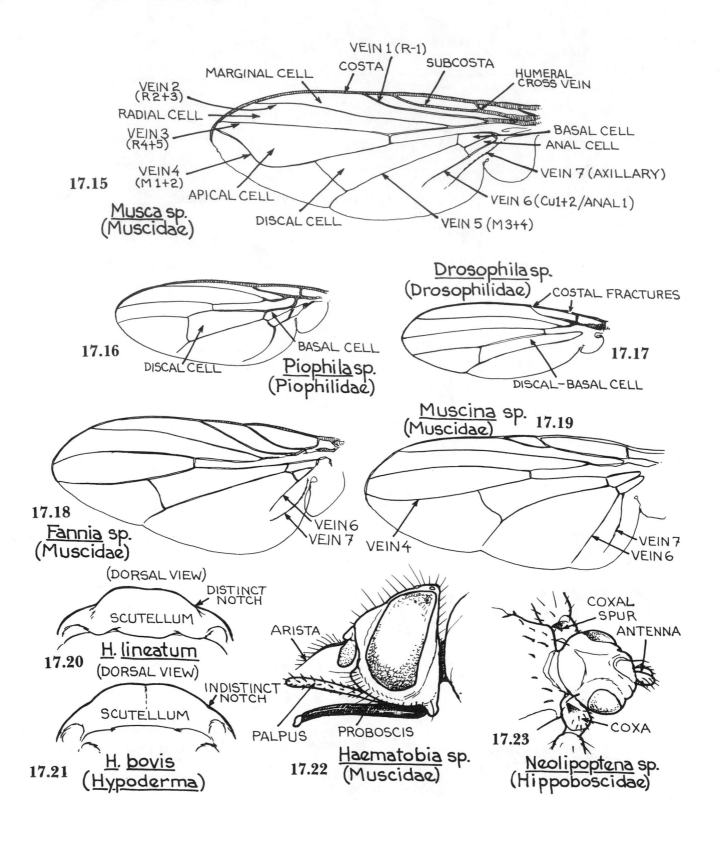

17.15 Musca sp. (Muscidae)

17.16 Piophila sp. (Piophilidae)

17.17 Drosophila sp. (Drosophilidae)

17.18 Fannia sp. (Muscidae)

17.19 Muscina sp. (Muscidae)

17.20 H. lineatum (DORSAL VIEW)

17.21 H. bovis (Hypoderma) (DORSAL VIEW)

17.22 Haematobia sp. (Muscidae)

17.23 Neolipoptena sp. (Hippoboscidae)

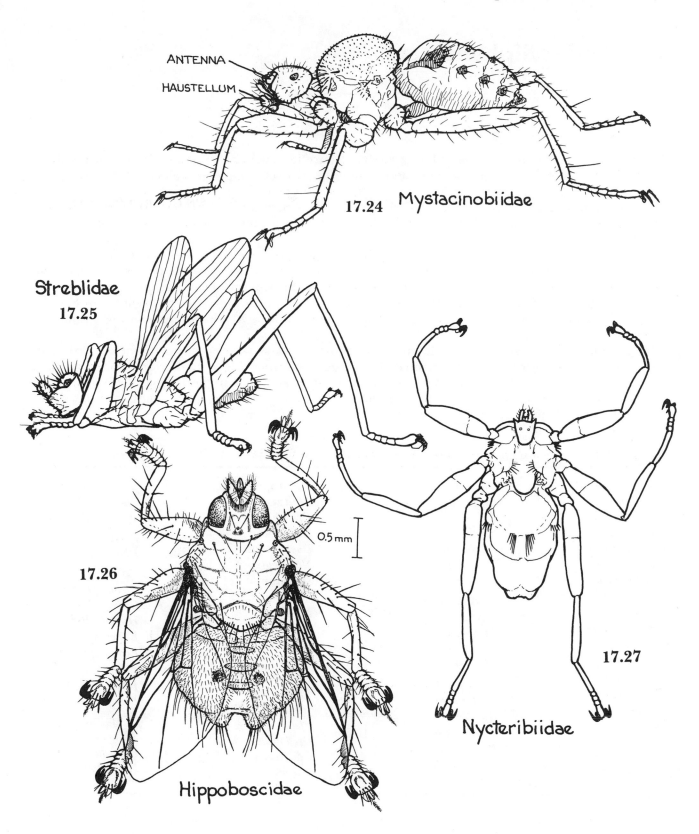

ANTENNA

HAUSTELLUM

17.24 Mystacinobiidae

Streblidae
17.25

0.5 mm

17.26

Hippoboscidae

17.27

Nycteribiidae

18   Order Siphonaptera

Study the general anatomy of a female flea as given in Figure 18.1, familiarizing yourself with the terminology employed. Note the prominent spermatheca and the relatively simple genitalia characteristic of the female flea. Compare with Figure 18.3, illustrating the complex genitalia of the male flea. The spermatheca of the female and much of the male genitalia are internal structures which stand out prominently after the specimens have been cleared of the softer body contents with a caustic solution such as potassium hydroxide.

In the male flea (Figure 18.3) note particularly the springlike penis rods and accessory apodemal rod, the spatulate aedeagal apodeme, the clasper with its dorsal fixed process and movable process or finger, the upper and ventral lobes of sternum IX, the external lobe of the aedeagus known as the crochet, and the modified tergite and sternite of segment VIII (T VIII, S VIII), which protect and sometimes enclose the genitalia. In both sexes the anus opens between small lobes (T X, S X) which represent the tergite and sternite of segment X.

Using the key given below, identify the specimens provided. Make sketches of significant key structures where these are not provided in the manual. A phylogenetic listing of genera of North American fleas according to family precedes the key. Once recognized, families of fleas are fairly easily discerned, but to provide characters of separation without exceptions is difficult. Therefore, genera have been keyed directly, and the list is provided for family assignment.

Phylogenetic listing of genera of North American fleas and families to which they belong

Hystrichopsyllidae	*Delotelis*	**Ceratophyllidae**	**Vermipsyllidae**
Stenoponia	*Megarthroglossus*	*Thrassis*	*Chaetopsylla*
Hystrichopsylla	*Callistopsyllus*	*Opisocrostis*	**Rhopalopsyllidae**
Atyphloceras	*Stenistomera*	*Oropsylla*	*Polygenis*
Rhadinopsylla	*Anomiopsyllus*	*Dactylopsylla*	*Rhopalopsyllus*
Paratyphloceras	*Jordanopsylla*	*Nosopsyllus*	**Pulicidae**
Trichopsylloides	*Conorhinopsylla*	*Diamanus*	*Ctenocephalides*
Corypsylla	**Ischnopsyllidae**	*Megabothris*	*Cediopsylla*
Nearctopsylla	*Sternopsylla*	*Malaraeus*	*Hoplopsyllus*
Corrodopsylla	*Myodopsylla*	*Monopsyllus*	*Euhoplopsyllus*
Doratopsylla	*Eptescopsylla*	*Ceratophyllus*	*Echidnophaga*
Ctenophthalmus	*Myodopsylloides*	*Dasypsyllus*	*Pulex*
Carteretta	**Leptopsyllidae**	*Mioctenopsylla*	*Juxtapulex*
Phalacropsylla	*Leptopsylla*	*Amphalius*	*Xenopsylla*
Neopsylla	*Amphipsylla*	*Jellisonia*	**Tungidae**
Tamiophila	*Ctenophyllus*	*Pleochaetis*	*Tunga*
Meringis	*Peromyscopsylla*	*Orchopeas*	*Rhynchopsyllus*
Epitedia	*Odontopsyllus*	*Opisodasys*	*Hectopsylla*
Catallagia	*Dolichopsyllus*	*Taropsylla*	

The combs (= ctenidia) of fleas are so obvious they are used in the first couplet of most keys for separating species of fleas. Unfortunately, groups thus separated do not contain phylogenetically related fleas, and the categories are not taxonomic. Combs on fleas occur independently in several unrelated groups of fleas. In several instances a few members bear no combs. In order to separate fleas easily and effectively, the following key relies on readily observable characters whenever advantageous. When taxonomic or "natural" divisions of groups are not achieved, the separating characters are termed "artificial."

Key to most genera and some common species of North American and other Siphonaptera[a]

1. With neither pronotal nor genal comb..2
   - With pronotal comb only...11
   - With both pronotal and genal comb..34

2. Sensilium (pygidium) (see Figure 18.1) with 8 pits on each side; inner side of hind coxa without spiniform bristles; antesensilial bristles absent; female fossae of spiracles of abdominal segments II–IV minute, V–VIII huge (jigger or chigoe flea – human pest in tropics)............................Tungidae, *Tunga penetrans*
   - Sensilium with 14 or more pits on each side; inner side of hind coxa with or without spiniform bristles (Figure 18.4); antesensilial bristles present or absent; in both sexes all fossae of spiracles of abdominal segments II–VIII same size (huge in *Chaetopsylla* (Figure 18.53), small to average in others) 3

3. Combined thoracic terga shorter than first abdominal tergum (Figure 18.7); labial palp soft, membranous, unsegmented; anterior margin of head angular; worldwide in warm climates, sessile on a variety of hosts (sticktight flea) ....*Echidnophaga gallinacea*
   - Combined thoracic terga longer than first abdominal tergum (Figures 18.1, 18.2); labial palps stiff, with 4 or more segments 4

4. Most abdominal terga with more than 1 row of setae (Figure 18.1); labial palps with 5 or more segments 5
   - Abdominal terga with 1 (or no) row of setae; labial palps with 4 or 5 segments...6

5. Antesensilial setae absent; fossae of abdominal spiracles funnel-shaped and extremely broad – more than 4 times as broad as atrium (Figure 18.53); labial palps with more than 5 segments; on Carnivora...........................Vermipsyllidae, *Chaetopsylla*
   - Antesensilial seta or setae present; fossae of abdominal spiracles tube-shaped and not broad – no more than 2 times as broad as atrium; labial palps of 5 segments; on Rodentia, particularly cotton rats in N. America ....Rhopalopsyllidae, *Polygenis gwyni*

6. Eye inapparent; labial palps with 4 segments; small fleas, males less than 2 mm, females less than 2.5 mm..*Anomiopsyllus*
   - Eye present, well developed; labial palps with 4 or 5 segments; medium-size fleas, males greater than 2.5 mm, females greater than 3.0 mm................................Pulicidae (in part) 7

7. Pleural ridge of mesothorax present (Figure 18.8); ocular bristle on level with eye; worldwide on *Rattus* 8
   - Pleural ridge of mesothorax absent (Figure 18.4); ocular bristle below eye; on a wide variety of mammals....................................10

8. Male antesensilial seta marginal and at apex of a process; setae of process 1 of clasper stout and one of these bent; outer (distal) arm of sternum IX rod-shaped and straight; female antesensilial seta not marginal; head of spermatheca broader than base of tail; an African species on *Rattus*, now widely distributed ....*Xenopsylla brasiliensis*
   - Male antesensilial seta not marginal, not on a process; setae of process 1 of clasper of average thickness or slender; outer arm of sternum IX paddle-shaped or curved; female antesensilial

seta nor marginal; head of spermatheca narrower than base of
tail 9

9. Male sternum IX curved, narrow; posterior margin
sclerotized; process 1 of clasper not broad; female tail of
spermatheca enlarged ventrally; on *Rattus* (domestic or
wild) ....*Xenopsylla astia*[b] and *vexabilis*[b]
   • Male sternum IX straight, broad toward apex; not sclerotized;
   process 1 of clasper broad; female ventral margin of tail and
   head about on an equal plane (Oriental rat flea – efficient
   transmitter of plague and murine typhus) ....*Xenopsylla cheopis*

10. Male dorsal aedeagal sclerite long and slender (Figure 18.9);
crochet expanded apically; worldwide on a wide variety of
mammals including man (so-called human flea) ....*Pulex irritans*
   • Male dorsal aedeagal sclerite broad throughout (Figure 18.10);
   crochet small and elongate; in N. America – on Carnivora,
   Rodentia, Artiodactyla, etc...*Pulex simulans*

11. Abdominal terga II to VI with single row of bristles 12
   • Abdominal terga II to VI with more than 1 row of bristles 16

12. Eyes well developed; lateral wall of mesocoxa without distinct
break, suture, or ridge (Figure 18.4); abdominal terga lacking
marginal spinelets ....Pulicidae (in part) 13
   • Eyes vestigial (Figure 18.26); lateral wall of mesocoxa with
   distinct break, suture, or ridge (Figures 18.33, 18.34);
   abdominal terga with marginal spinelets (Figure 18.1) ....Hystrichopsyllidae (in part) 14

13. Male: process 1 of clasper divided into two lobes by deep sinus;
female tail of spermatheca curved at apex, head globular to
oval; segment 3 (clava) of antenna with distinct
pseudosegments on ventral side; on *Lepus* and *Sylvilagus*
(hares and rabbits) ....*Euhoplopsyllus glacialis*
   • Male process 1 of clasper oval, female tail of spermatheca
   broadly curved, head oval; segment 3 (clava) of antenna fused
   on ventral side (as it lies in fossa); on *Spermophilus* (ground
   squirrels) ....*Hoplopsyllus anomalus*

14. Four to 5 rows of preantennal bristles; fracticipit (Figure
18.26)[c]; abdominal sterna with bristles spiniform (at least in
basal portion of bristle ....*Stenistomera*
   • One to 2 rows of preantennal bristles; integricipit; abdominal
   sterna with bristles appearing slender, as is typical of most
   fleas 15

15. Labial palps 4-segmented; segment 2 (pedicel) of antenna
forming a sheath around base of segment 3 (clava) ....*Callistopysllus*
   • Labial palps 5-segmented; pedicel of antenna not as above ....*Megarthroglossus*

16. Anterior-most row of preantennal bristles located anterior to
(Figure 18.27) or directly over (Figure 18.47) sclerified parts of
cibarial pump; head bristles usually numerous 17
   • Preantennal bristles (if existing) all posterior to cibarial pump;
   head bristles usually scanty, seldom arranged in rows of 4 or
   more bristles (Figure 18.48) ....Ceratophyllidae (most) 21

17. Arch of tentorium visible in front of or above eye (Figures
18.27, 18.28); spiniform bristles (setae) on metacoxa present
(Figure 18.28) or absent; if absent, anterior-most rows of

preantennal bristles distinctly anterior to cibarial pump
(Figure 18.27) 18
- Arch of tentorium not visible (Figures 18.47, 18.48); metacoxa
  without patch of spiniform bristles; anterior-most row of
  preantennal bristles over cibarial pump (Figure 18.47) . . . . .Ceratophyllidae 21

18. Metanotum without marginal (apical) spinelets; trabecula
    centralis absent; on rodents and occasionally insectivores . . . . .Hystrichopsyllidae (in part) 19
- Metanotum with marginal spinelets; trabecula centralis
  present (Figure 18.28); on lagomorphs . . . . .Leptopsyllidae (in part) 20

19. Four or more apical spinelets on most abdominal terga;
    metacoxa lacking spiniform bristles; female spermatheca
    double . . . .*Atyphloceras*
- Fewer than 4 apical spinelets on each side of abdominal terga;
  metacoxa with row or patch of spiniform bristles (see Figure
  18.24) . . . .*Catallagia*

20. Preantennal region with a row of pigmented spiniform
    bristles; metacoxa without spiniform bristles; on *Ochotona*
    (cony) . . . .*Ctenophyllus terribilis*
- Preantennal region with simple bristles; metacoxa with a patch
  of spiniform bristles (Figure 18.28); on *Lepus* and *Sylvilagus*
  (hares and rabbits) . . . .*Odontopsyllus*

21. Eyes vestigial and unpigmented; on Geomyidae (pocket
    gopher) . . . .*Dactylopsylla*
- Eyes of medium to large size, pigmented 22

22. A number of small lateral bristles on fore femur (Figure
    18.29) 23
- One or no lateral bristles on fore femur (Figure 18.30) 33

23. Total of 24 or more teeth in pronotal comb; bird fleas 24
- Fewer than 24 teeth in pronotal comb; mammal fleas 25

24. Third pair of plantar bristles on tarsi V shifted ventrally
    (Figure 18.31); on birds and occasionally mammals . . . .*Dasypsyllus gallinule*
- Plantar bristles all lateral . . . .*Ceratophyllus*

25. Anterior inner surface of meso- and metacoxa with long thin
    bristles from base to apex, aside from those along anterior
    margin of coxa (Figure 18.33); labial palps often exceeding
    forecoxa and trochanter in length 26
- Basal part of anterior inner surface of meso- and metacoxa
  with no bristles except those along anterior margin (Figure
  18.34); labial palps seldom exceeding forecoxa and trochanter
  in length 29

26. Basal abdominal sternum (S. II) with patch of lateral setae
    (Figure 18.32); on *Cynomys* and *Spermophilus* (prairie dogs and
    ground squirrel) . . . .*Opisocrostis*
- Basal abdominal sternum without patch of setae; often on
  Spermophilus, but rarely on *Cynomys* 27

27. Male sternum VIII much reduced, without bristles (Figure
    18.35); movable process (finger) curved, slender; female
    spermatheca (Figure 18.36) head globular, tail usually
    constricted at point of attachment to head; stylet (Figure
    18.37) with very small dorsal bristle . . . .*Diamanus montanus*

- Male sternum VIII not highly modified by reduction, usually rounded or produced apically, covered with bristles; movable process curved, broad (Figure 18.38); female spermatheca head globular, oblong, or pyriform, tail not constricted (Figures 18.39, 18.42); stylet usually with long dorsal bristle (Figure 18.40) 28

28. Male sternum VIII narrow, bearing long apical bristles (Figure 18.38); female head of spermatheca longer than broad, pyriform shape (Figure 18.39) ....*Oropsylla*
   - Male sternum VIII broad, bearing several to many bristles (Figure 18.41); female head of spermatheca broader than long (Figure 18.42) ....*Thrassis*

29. Male sternum VIII vestigial, no bristles (Figure 18.43); penis rods long and coiled; female tail of spermatheca curved around head (Figure 18.44); apex of bursa copulatrix rolled up as a spiral; worldwide, temperate, on *Rattus* (European rat flea) ....*Nosopsyllus fasciatus*
   - Male sternum VIII reduced or long, narrow, often with bristles; penis rods shorter and coiled less than 1.5 revolutions; female tail of spermatheca not curved around head; apex of bursa copulatrix not as above; on cricetid rodents 30

30. Fossa of spiracular opening of tergum VIII enlarged (Figure 18.45) ....*Megabothris*
   - Fossa of spiracular opening of tergum VIII not enlarged (Figure 18.46) 31

31. Occiput (postantennal region) with 2 or more long bristles behind base of antennal groove and 3 or more in middle, arranged in rows (Figure 18.47) ....*Pleochaetis*
   - Occiput with 1 or no long bristles behind base of antennal groove (Figure 18.48) 32

32. Eye reduced, size as small as or smaller than shown in Figure 18.48, its longitudinal diameter shorter than the distance of the eye from apex of incrassate portion of genal lobe ....*Malaraeus*
   - Eye not reduced (see Figure 18.47) ....*Monopsyllus*

33. Male movable process with 4 to 7 short, equal spiniforms directed upward (Figure 18.49); female ventral margin of sternum X (anal) angular near middle; head of spermatheca barrel-shaped (Figure 18.50) ....*Orchopeas*
   - Male movable process with 2 or 3 medium to long, unequal spiniforms directed downward or distad (Figure 18.51); female ventral margin of sternum X not angular; head of spermatheca more convex dorsally than ventrally (Figure 18.52) ....*Opisodasys*

34. Eye present, well developed; abdominal terga II–VI with single row of bristles; metanotum and abdominal terga without marginal spinelets; lateral surface of mesocoxa without break (Figure 18.4) ....Pulicidae (part) 35
   - Eye inapparent or, if present, not well developed; abdominal terga II–VI with more than 1 row of bristles; abdominal terga with marginal spinelets; lateral surface of mesocoxa with suture or break (Figure 18.33, 18.34) ....Hystrichopsyllidae (in part), Leptopsyllidae (in part), Ischnopsyllidae 37

35. Genal comb oblique with straight, blunt teeth, on *Lepus* and
Sylvilagus (hare and rabbits) ....*Cediopsylla*
    • Genal comb horizontal with curved, sharp teeth (Figures 18.5,
18.6); on Carnivora 36

36. Anterior-most pair of genal spines nearly same length as next
pair of genal spines; anterior margin of head low (Figure
18.5); dorsal border of hind tibia with 5 to 6 notches, with one
notch bearing a stout spine between the postmedian notch and
the apex (Figure 18.5); worldwide on dogs and cats ....*Ctenocephalides felis*
    • Anterior-most pair of genal spines shorter than next pair of
genal spines; anterior margin of head high (Figure 18.6);
dorsal border of hind tibia with 7 to 8 notches, with 2 notches
each bearing a stout spine between the postmedian notch and
the apex (Figure 18.6); worldwide in warm areas on dogs and
cats ....*Ctenophalides canis*

37. Abdominal tergum I with comb; on field rodents...*Stenoponia*
    • No abdominal comb 38

38. Genal comb of 4 or more teeth 39
    • Genal comb of fewer than 4 teeth 45

39. Genal comb anterior to eye position; teeth spatulate; pronotal
comb curved (Figure 18.11) 40
    • Genal comb anterior to or ventral to eye position; teeth slender
(Figure 18.12); pronotal comb straight (as in Figure 18.1) 41

40. Genal comb of 5 teeth; dorsal portion of abdominal terga not
sclerotized ....*Nearctopsylla*
    • Genal comb of 6 teeth; dorsal portion of abdominal terga
sclerotized and darkened ....*Corypsylla*

41. Two spiniform bristles along frontal margin; clypeal area long
and rounded (Figure 18.12); on *Mus* (house mice) and
sometimes *Rattus* ....*Leptopsylla segnis*
    • No spiniform bristles along frontal margin; clypeal area not
rounded 42

42. Genal comb of 4 teeth arranged horizontally (Figure 18.13);
basal pair of plantar bristles of tarsal segment V medially
displaced, with 4 lateral bristles (Figure 18.14); on Insectivora
(especially shrews) 43
    • Genal comb of 5 or more teeth (very rarely 4) arranged
horizontally to almost vertically; all plantar bristles of segment
V lateral 44

43. Genal comb extends posteriorly and dorsally so that the
uppermost tooth conceals genal process beneath; apical
margins of abdominal terga uneven ....*Corrodopsylla curvata*
    • Last genal tooth ventral to genal process, which is clearly
visible; apical margins smooth ....*Doratopsylla blarinae*

44. Male sternum IX lacking heavy spiniforms (Figure 18.15);
female spermatheca single; genal teeth 4 to 6; hosts often
sciurid rodents ....*Rhadinopsylla*
    • Male sternum IX with heavy spinelike bristles along posterior
border of external arm (Figure 18.17); female spermatheca
double; large fleas, over 3 mm long; genal teeth 5 to 12 (Figure
18.16) usually on cricetid rodents, rarely on Insectivora ....*Hystrichopsylla*

45. "Genal" comb preoral, located at extreme anterior end of the ventral margin of the head, teeth broad and flat (Figures 18.18, 18.19); on Chiroptera (bat fleas) .... *Ischnopsyllidae* 46
    - Genal comb posterior to mouthparts, teeth not flat (Figures 18.20, 18.21); not on Chiroptera 47

46. Apex of maxilla squarely truncate (Figure 18.18); abdominal segments I–VII with "false combs" (enlarged bristles) .... *Myodopsylla*
    - Apex of maxilla truncate, but with one angle extended (Figure 18.19); no "combs" on abdomen .... *Sternopsylla distincta*

47. Genal comb with 3 teeth 48
    - Genal comb with 2 teeth 49

48. Two anterior teeth of genal comb overlapping .... *Carterella carteri*
    - Two anterior teeth of genal comb not overlapping (Figure 18.20) .... *Ctenophthalmus pseudagyrtes*

49. Genal teeth separate, not overlapping; clypeal region (anterior margin) of head with short spinelike bristles; clypeal margin long and rounded (Figure 18.21); on cricetid rodents .... *Peromyscopsylla*
    - Genal teeth overlapping (Figures 18.23, 18.25); clypeal region without short spinelike bristles; clypeal margin not as above 50

50. Clypeal notch present (Figure 18.23); metatarsus V with but 4 lateral plantar bristles (Figure 18.24b) (rare exceptions, Figure 18.24c); marginal abdominal spinelets prominent; row of spinelike bristles on inner surface of hind coxa absent or present (Figure 18.24d) 51
    - Clypeal notch absent (Figures 18.1, 18.25); metatarsus V with 4 lateral plantar bristles and 1 medially displaced pair; marginal abdominal spinelets barely discernible (Figure 18.1) or absent; row of spiniforms present on inner surface of metacoxa (see Figure 18.24d) 53

51. All plantar bristles of all tarsi V lateral; metacoxa without row of spinelike bristles on inner surface (Figure 18.22); on sciurid rodents 52
    - Fore and mid tarsi with 4 lateral pairs and a basal medially placed pair of plantar bristles; hind tarsus with only 4 lateral pairs (Figure 18.24b); row of spinelike bristles on hind coxa (Figure 18.24d); on cricetid rodents and shrew .... *Epitedia*

52. Basal abdominal sternum with setae along ventral margin; large fleas over 4 mm long; on *Tamias* (eastern chipmunk) .... *Tamiophila grandis*
    - Basal abdominal sternum without ventral setae; fleas less than 3 mm long; on *Spermophilus* (ground squirrels) .... *Neopsylla inopina*

53. Labial palps equal length of fore coxa (Figure 18.1); marginal spinelets on abdominal terga barely discernible (Figure 18.1); often on *Neotoma* .... *Phalacropsylla*
    - Labial palps 0.75 length of fore coxa (Figure 18.25); no marginal spinelets on abdominal terga; often on *Dipodomys* .... *Meringis* 54

54. Male sternum IX not bilobed; apex of distal arm acuminate or blunt; female head of spermatheca as broad posteriorly as anteriorly and may be constricted in middle .... *Meringis arachis* group (12 species)

- Male sternum IX divided into dorsal and ventral lobes; apex of dorsal portion rounded; female head of spermatheca broad basally (where tail is attached); narrow apically . . . .*Meringis parkeri* group (5 species)

[a]Prepared by Harold E. Stark.
[b]Distribution allopatric.
[c]Antennal suture across top of head, suggesting a broken-head appearance.

References

Barnes, A. M., V. J. Tipton, and A. J. Wildie. 1977. The subfamily Anomiopsyllinae (Hystrichopsyllidae: Siphonaptera). I. A revision of the genus *Anomiopsyllus* Baker. *Gt. Basin Nat. 37*:138–206.

Holland, G. P. 1949. *The Siphonaptera of Canada.* Dominion of Canada, Dept. of Agriculture, Ottawa, Publ. 817, Tech. Bull. 70. 306 pp.

Hopkins, G. H. E., and M. Rothschild. 1953–72. *Catalogue of the Rothschild Collection of Fleas*, vols. 1–5. Cambridge University Press and British Museum (Natural History), London.

Hopla, C. E. 1980. A study of the host relations and zoogeography of *Pulex.* In *Proc. Int. Conf. on Fleas,* ed. R. Traub and H. Starcke. A. A. Balkema, Rotterdam, pp. 185–208.

Johnson, P. T. 1957. A Classification of the Siphonaptera of South America. *Mem. Entomol. Soc. Wash. 5*:1–299.

– 1961. *A revision of the species of* Monopsyllus *Kolenati in North America (Siphonaptera, Ceratophyllidae).* U.S. Dept. of Agriculture Tech. Bull. 1227, pp. 1–69.

Johnson, P. T., and R. Traub. 1954. *Revision of the Flea Genus* Peromyscopsylla. Smithsonian Misc. Collection 123(4), pp. 1–68.

Rothschild, M., and R. Traub. 1971. A revised glossary of terms used in the taxonomy and morphology of fleas. In *Catalogue of the Rothschild Collection of Fleas,* vol. 5, ed. G. Hopkins and M. Rothschild. Cambridge University Press and British Museum (National History), London, pp. 8–85.

Smit. F. G. A. M. 1954. Identification of fleas, Annex 2. In *Plague,* ed. R. Pollitzer. WHO Monograph No. 22, WHO, Geneva, pp. 648–82.

– 1970. Siphonaptera. In *Taxonomist's Glossary of Genitalia in Insects,* ed. S. L. Tuxen. Monksgaard, Copenhagen, pp. 141–55.

Snodgrass, R. E. 1946. *The Skeletal Anatomy of Fleas (Siphonaptera).* Smithsonian Misc. Collection 104(18), pp 1–89. 21 plates.

Stark, H. E. 1958. *The Siphonaptera of Utah.* Communicable Disease Center, Public Health Service, Atlanta. xiii + 239 pp.

– 1970. *A Revision of the Flea Genus* Thrassis *Jordan 1933 (Siphonaptera: Ceratophyllidae),* University of California Publications in Entomology, vol. 53. University of California Press, Berkeley. 184 pp.

Tipton, V. J. and E. Mendez. 1966. The fleas (Siphonaptera) of Panama. In *Ectoparasites of Panama,* ed. R. Wenzel and V. J. Tipton. Field Museum of Natural History, Chicago, pp. 289–386.

Tipton, V. J., H. E. Stark, and J. A. Wildie. 1979. Anomiopsyllinae (Siphonaptera, Hystrichopsyllidae), II. The genera *Callistopsyllus, Conorhinopsylla, Megarthroglossus* and *Stenistomera. Gt. Basin Nat. 38*:351–418.

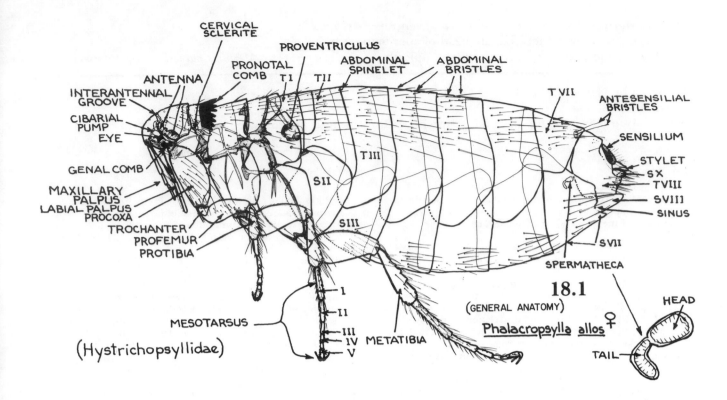

INTERANTENNAL GROOVE

CIBARIAL PUMP

EYE

ANTENNA

GENAL COMB

MAXILLARY PALPUS

LABIAL PALPUS

PROCOXA

TROCHANTER

PROFEMUR

PROTIBIA

CERVICAL SCLERITE

PRONOTAL COMB

PROVENTRICULUS

ABDOMINAL SPINELET

ABDOMINAL BRISTLES

T I T II

T III

S II

S III

T VII

ANTESENSILIAL BRISTLES

SENSILIUM

STYLET

S X

T VIII

S VIII

SINUS

S VII

SPERMATHECA

MESOTARSUS

I

II

III

IV

V

METATIBIA

METATIBIA

(Hystrichopsyllidae)

18.1
(GENERAL ANATOMY)

Phalacropsylla allos ♀

HEAD

TAIL

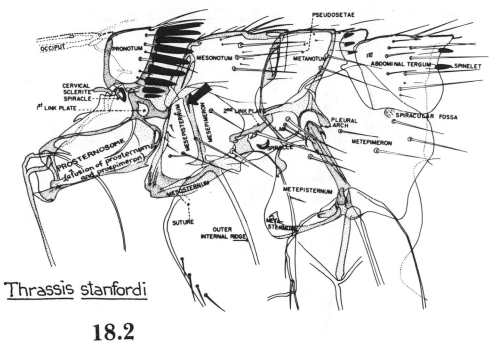

PSEUDOSETAE

OCCIPUT
PRONOTUM
MESONOTUM
METANOTUM
1ST ABDOMINAL TERGUM
SPINELET
CERVICAL SCLERITE
SPIRACLE
1st LINK PLATE
2nd LINK PLATE
SPIRACULAR FOSSA
PROSTERNOSOME (a fusion of prosternum and proepimeron)
MESEPISTERNUM
MESEPIMERON
PLEURAL ARCH
SPIRACLE
METEPIMERON
MESOSTERNUM
METEPISTERNUM
SUTURE
OUTER INTERNAL RIDGE
META-STERNUM

Thrassis stanfordi

18.2

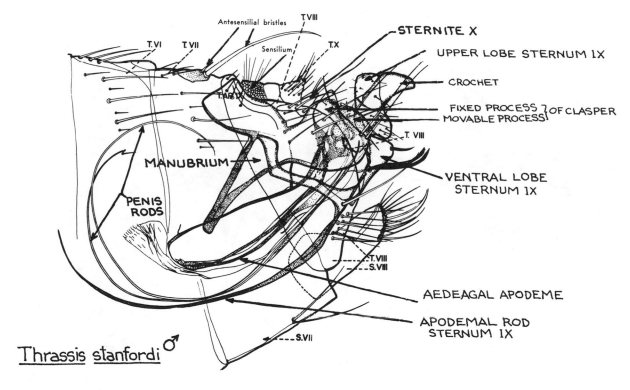

Antesensilial bristles   T.VIII
T.VI   T.VII   Sensilium   T.X
T ABD IX
STERNITE X
UPPER LOBE STERNUM IX
CROCHET
FIXED PROCESS ⎤ OF CLASPER
MOVABLE PROCESS ⎦
T. VIII
MANUBRIUM
PENIS RODS
VENTRAL LOBE STERNUM IX
T. VIII
S. VIII
AEDEAGAL APODEME
APODEMAL ROD STERNUM IX
S. VII

Thrassis stanfordi ♂

18.3

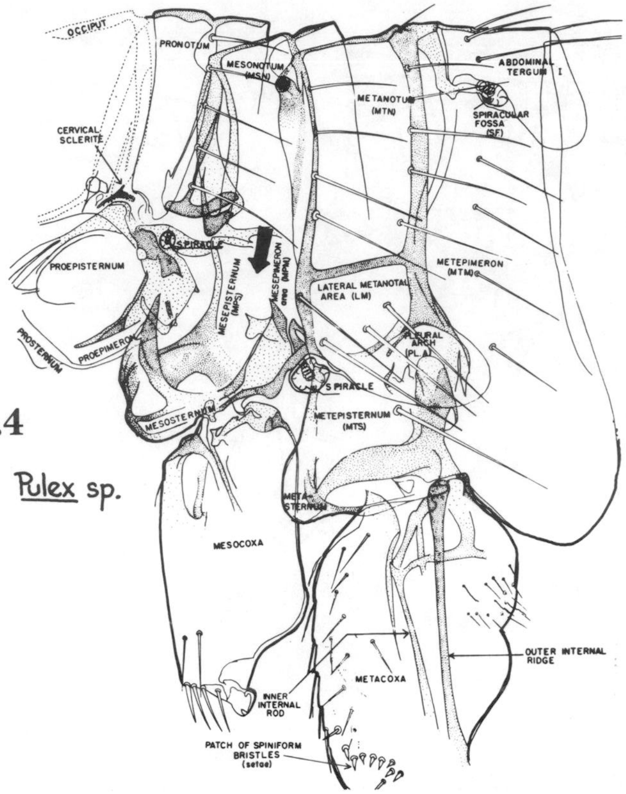

OCCIPUT
PRONOTUM
MESONOTUM (MSN)
METANOTUM (MTN)
ABDOMINAL TERGUM I
SPIRACULAR FOSSA (SF)
CERVICAL SCLERITE
SPIRACLE
PROEPISTERNUM
MESEPISTERNUM (MPS)
MESEPIMERON (MPM)
LATERAL METANOTAL AREA (LM)
METEPIMERON (MTM)
PLEURAL ARCH (PL A)
PROSTERNUM
PROEPIMERON
SPIRACLE
METEPISTERNUM (MTS)
MESOSTERNUM
META-STERNUM
Pulex sp.
MESOCOXA
OUTER INTERNAL RIDGE
METACOXA
INNER INTERNAL ROD
PATCH OF SPINIFORM BRISTLES (setae)

18.4

HIND TIBIA
C. felis
POST MEDIAN NOTCH
SPINED NOTCH

18.5

C. canis
HIND TIBIA
POST MEDIAN NOTCH
SPINED NOTCHES

18.6

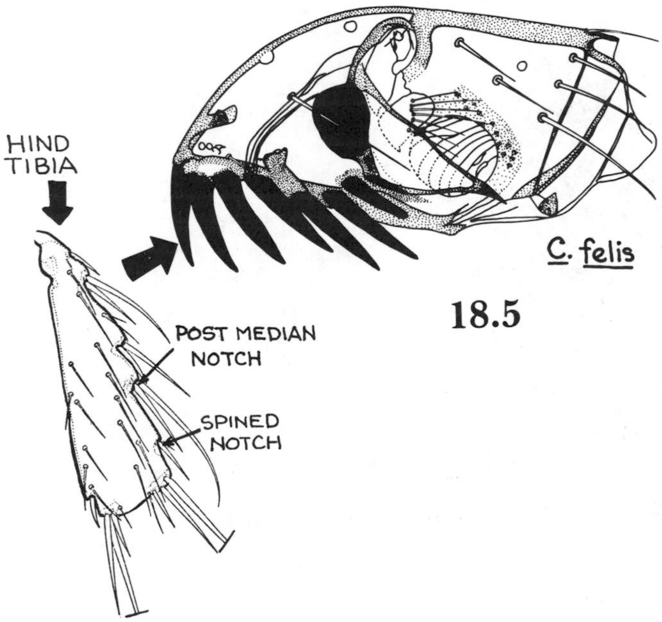

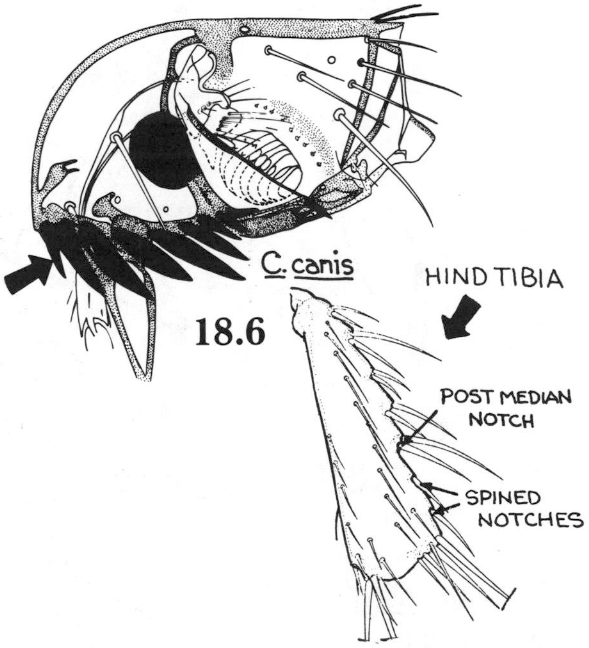

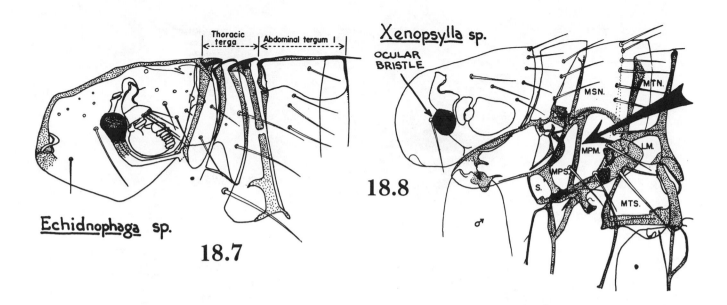

Thoracic terga | Abdominal tergum I

Echidnophaga sp.

18.7

Xenopsylla sp.

OCULAR BRISTLE

MSN. MTN. MPM. LM. MPS. S. MTS.

♂

18.8

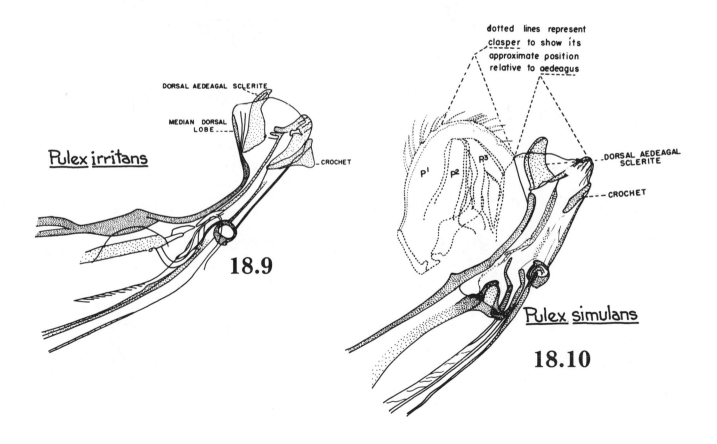

Pulex irritans

DORSAL AEDEAGAL SCLERITE

MEDIAN DORSAL LOBE

CROCHET

18.9

dotted lines represent clasper to show its approximate position relative to aedeagus

P¹ P² R³

DORSAL AEDEAGAL SCLERITE

CROCHET

Pulex simulans

18.10

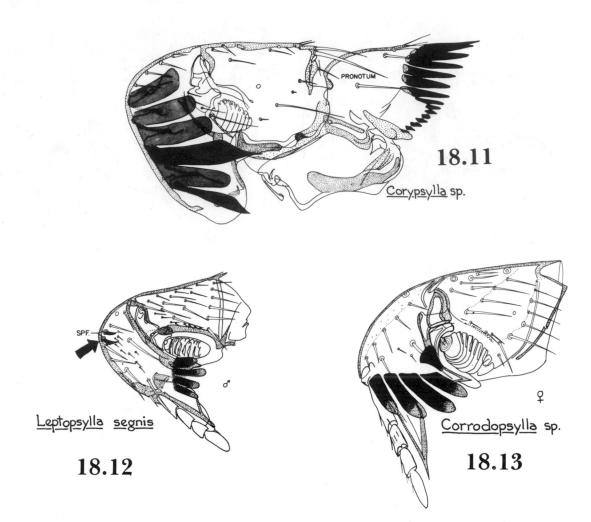

18.11

Corypsylla sp.

PRONOTUM

SPF.

Leptopsylla segnis

♂

18.12

Corrodopsylla sp.

♀

18.13

18.14

♂

Rhadinopsylla sp.

18.15

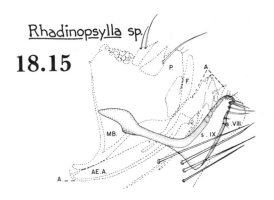

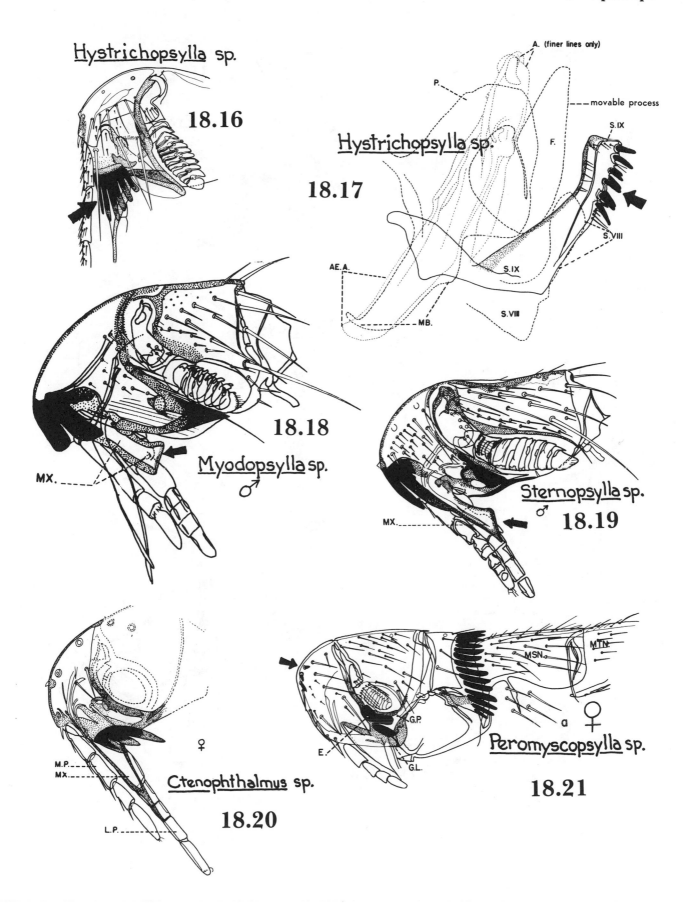

Hystrichopsylla sp.
18.16

Hystrichopsylla sp.
18.17

A. (finer lines only)
P.
movable process
F.
S.IX
AE.A.
S.IX
MB.
S.VIII
S.VIII

18.18
Myodopsylla sp. ♂
MX.

Sternopsylla sp. ♂
18.19
MX.

Ctenophthalmus sp. ♀
18.20
M.P.
MX.
L.P.

Peromyscopsylla sp. ♀
18.21
MTN.
MSN.
G.P.
E.
G.L.
a

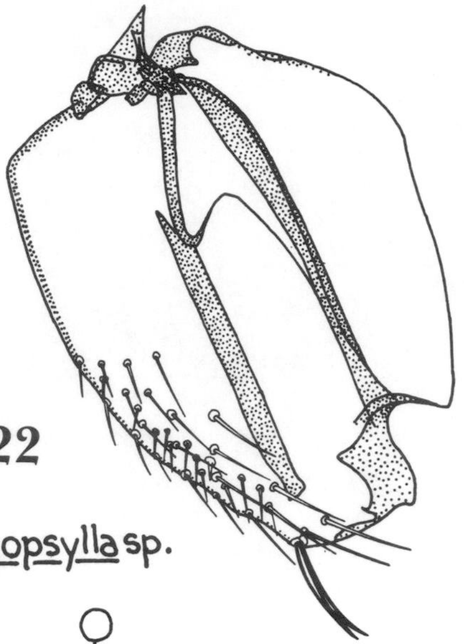

18.22

Neopsylla sp.

♀

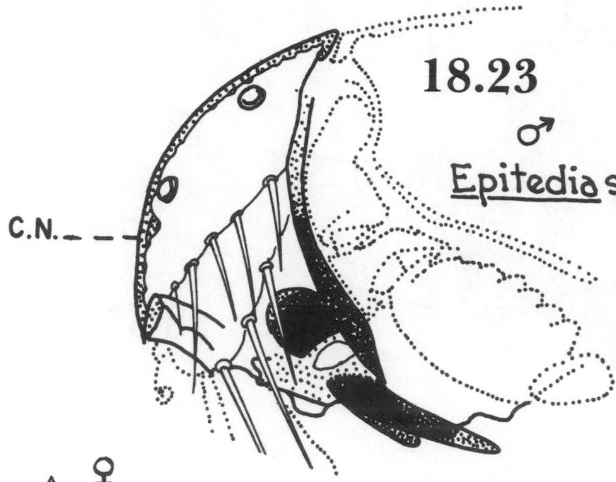

18.23

♂

Epitedia sp.

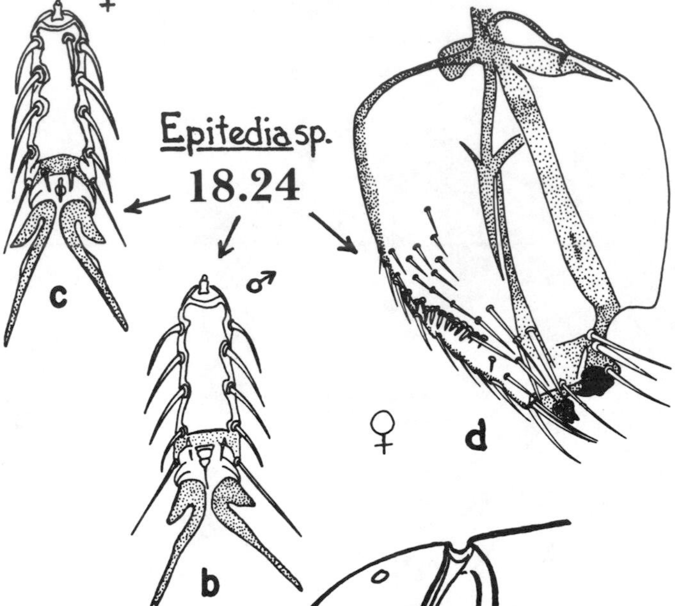

C.N.

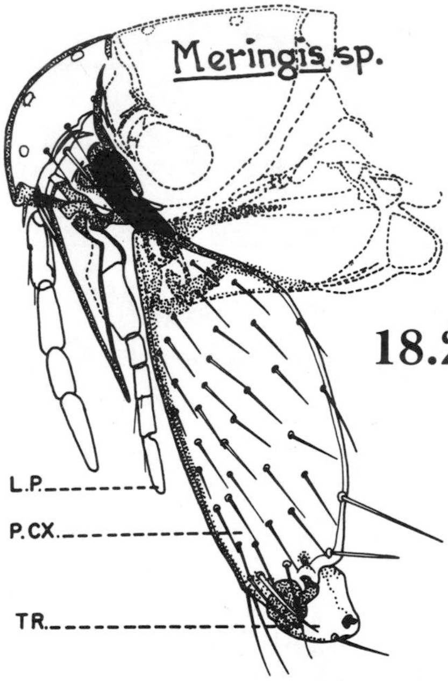

Meringis sp.

♀

18.25

L.P.

P.CX.

TR.

♀

Epitedia sp.

18.24

c

♂

b

♀

d

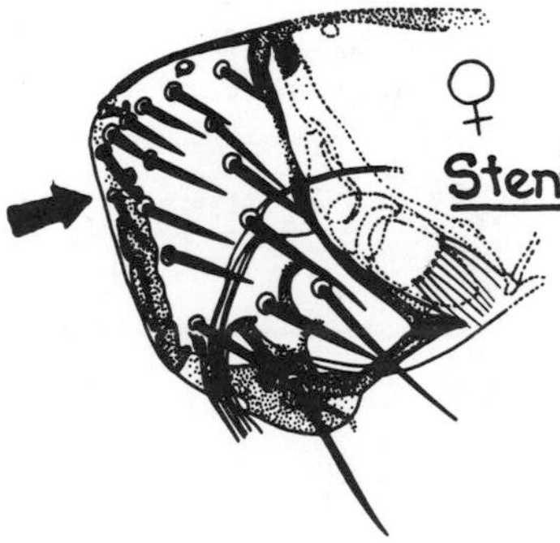

Stenistomera sp.

♀

18.26

♂

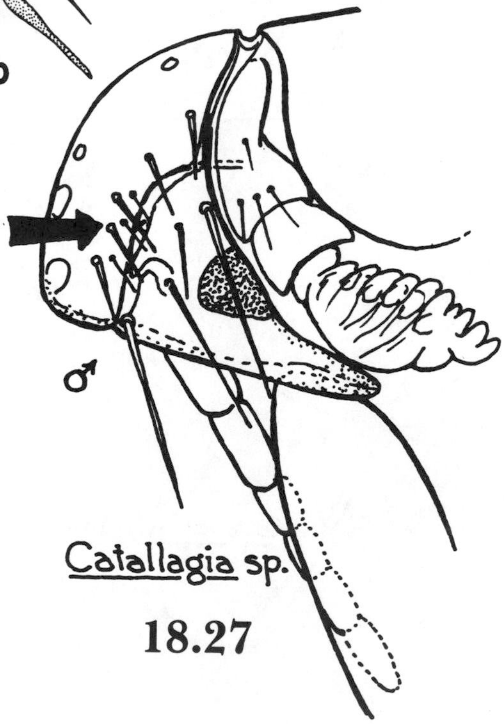

Catallagia sp.

18.27

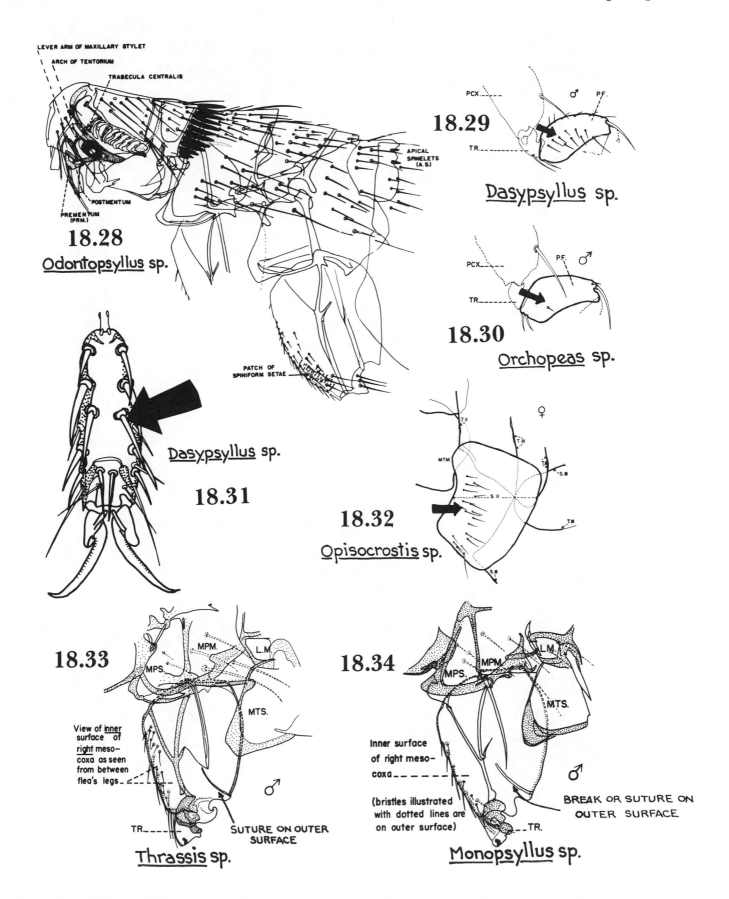

LEVER ARM OF MAXILLARY STYLET

ARCH OF TENTORIUM

TRABECULA CENTRALIS

APICAL
SPINELETS
(A. S.)

POSTMENTUM

PREMENTUM
(PRM.)

18.28

Odontopsyllus sp.

PATCH OF
SPINIFORM SETAE

PCX. ♂ P.F.

TR.

18.29

Dasypsyllus sp.

PCX. P.F. ♂

TR.

18.30

Orchopeas sp.

Dasypsyllus sp.

18.31

♀

T. II

T. III

MTM.

T. I

S. III

S. II

T. III

18.32

Opisocrostis sp.

S. III

18.33

MPM. L.M.

MPS.

MTS.

View of <u>inner</u>
surface of
right meso-
coxa as seen
from between
flea's legs

TR.

SUTURE ON OUTER
SURFACE

♂

Thrassis sp.

18.34

MPM. L.M.

MPS.

MTS.

Inner surface
of right meso-
coxa

(bristles illustrated
with dotted lines are
on outer surface)

TR.

BREAK OR SUTURE ON
OUTER SURFACE

♂

Monopsyllus sp.

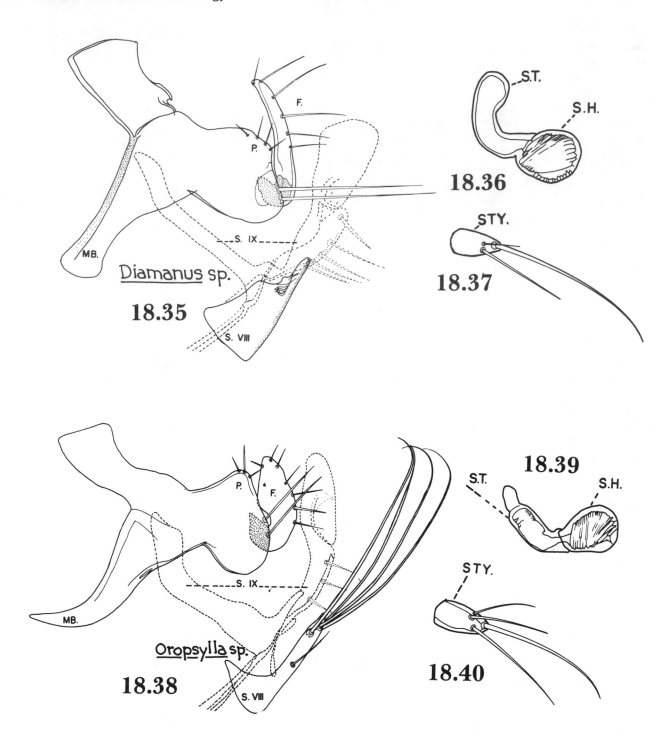

F.

P.

MB.

S. IX

S. VIII

Diamanus sp.

18.35

S.T.

S.H.

18.36

STY.

18.37

P.

F.

MB.

S. IX

S. VIII

Oropsylla sp.

18.38

18.39

S.T.

S.H.

STY.

18.40

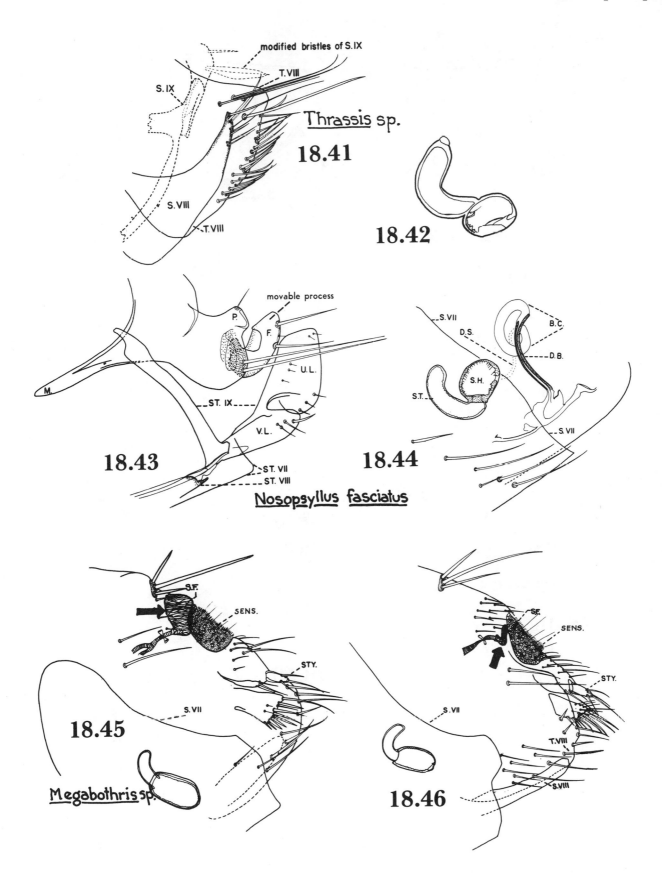

modified bristles of S.IX

T.VIII

S.IX

Thrassis sp.

18.41

S.VIII

T.VIII

18.42

movable process

P.

F.

M.

U.L.

ST. IX

V.L.

ST. VII

ST. VIII

18.43

Nosopsyllus fasciatus

S.VII

D.S.

B.C.

D.B.

S.H.

S.T.

S.VII

18.44

S.F.

SENS.

STY.

S.VII

T.VIII

18.45

Megabothris sp.

S.F.

SENS.

STY.

S.VII

T.VIII

S.VIII

18.46

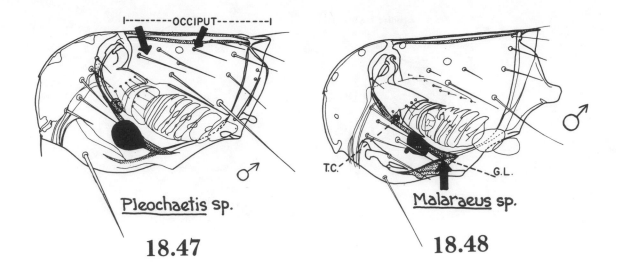

Pleochaetis sp.

18.47

Malaraeus sp.

18.48

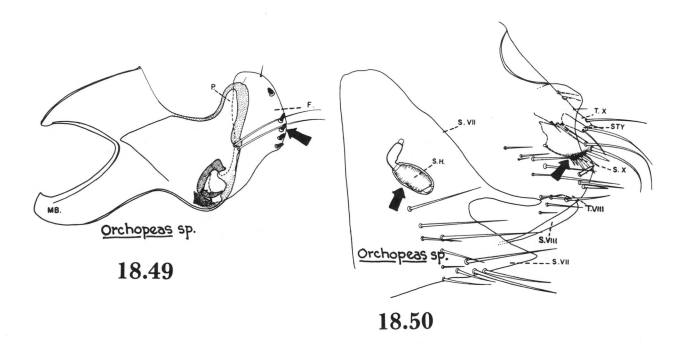

Orchopeas sp.

18.49

Orchopeas sp.

18.50

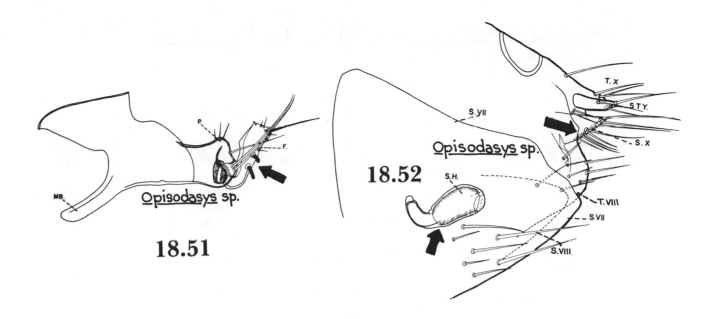

18.51

18.52

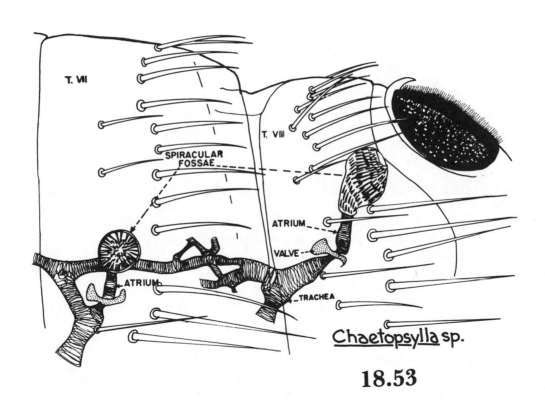

18.53

19 Order Parasitiformes: suborder Ixodida (ticks)

The subclass Acari of the class Arachnida is composed of several groups of arthropods that are known commonly as mites and ticks. Modern authorities differ as to the higher classification of these organisms. The classification followed here is patterned for convenience after that of Krantz (1978). This divides the Acari into two orders, Parasitiformes and Acariformes. Within the Parasitiformes there are included 4 suborders, Opilioacarida, Holothyrida, Gamasida, and Ixodida. Two of these suborders, Gamasida and Ixodida, contain many important parasites of man and other animals. The Ixodida (Metastigmata or Ixodoidea of authors) are known commonly as ticks. The Gamasida (Mesostigmata of authors) include many external and internal parasitic mites of a wide variety of vertebrate hosts as described in Exercise 20. The order Acariformes is divided into three suborders, Actinedida (Prostigmata of authors), Acaridida (Astigmata of authors), and Oribatida (Cryptostigmata of authors). Each of these suborders contains members important to the health of man and animals (see Exercise 20).

Members of the Ixodida possess lateral stigmatal openings behind coxae IV or laterally above coxae II–III. The prosoma, or appendage-bearing section, is completely fused, without any trace of segmentation to the opisthosoma, or abdominal section, behind the legs. Ticks are distinguished from gamasid mites by their relatively large size, by lack of elongate peritremes, by the possession of a true Haller's organ on the first pair of legs (Figure 19.17), by dentate faces of cheliceral digits directed externally (laterally), and by the presence of a well-developed hypostome provided ventrally with recurved teeth (Figure 19.10). Ticks are blood-sucking, obligatory ectoparasites.

Before using the key for identification of ticks, it is essential to determine if the specimen is a larva, nymph, or adult. Larvae have only three pairs of legs, in contrast to the four pairs seen in nymphs and adults. Nymphs lack the genital aperture (Figure 19.2), which occurs in the adult in the ventral area between the second to fourth pairs of coxae. Nymphs of the family Ixodidae have a scutum of the adult female type (Figure 19.1), but the basis capituli lacks the porose areas seen only in adult females (Figure 19.3) of ixodid, or hard, ticks.

Key to families (excluding Nuttalliellidae), selected genera, and species of Ixodida, adults

1. Capitulum situated anteriorly, not concealed from above; dorsal scutum present, spiracles posterior to coxae IV; well marked sexual dimorphism (Figures 19.1, 19.2) ....Ixodidae 2
   - Capitulum ventral, partially or completely concealed from above; dorsal scutum absent; spiracles usually anterior to coxae IV; sexual dimorphism slight (Figures 19.18, 19.19) ....Argasidae 18

Ixodidae

2. Anal groove distinct or indistinct, but never surrounding the anus anteriorly; festoons often present (Figure 19.2) 3
   - Anal groove distinct, surrounding the anus anteriorly (Figure 19.7); festoons absent (grooves and festoons best seen in unfed ticks) ....*Ixodes*
   (*Ixodes pacificus* the most commonly encountered tick of this genus in the Pacific Coast states) (see Keirans and Clifford, 1978, for keys and figures of North American species)

3. Capitulum short in relation to width; 2nd palpal segment not especially longer than wide (Figures 19.3, 19.6) 4
   - Capitulum long in relation to width; 2nd segment longer than wide (Figure 19.4)...10

4. Second palpal segment not acutely laterally produced at base
   (Figure 19.3); eyes present (Figure 19.1) 5
   ● Second palpal segment acutely laterally produced, extending
   beyond the basis capituli (Figure 19.6); eyes absent ....*Haemaphysalis*
   (Common on rabbits) ....*Haemaphysalis leporispalustris*)

5. Festoons absent; anal grooves faint or lacking 6
   ● Festoons present; anal grooves very distinct to indistinct
   (Figure 19.2)..7

6. Palpi compressed with acute transverse ridges dorsally and
   laterally (Figure 19.8); males with normal legs, adanal and
   accessory plates but without preanal plate ....*Boophilus*
   (Three species; primarily cattle parasites; *B. annulatus* in
   N. America)
   ● Palpi not ridged and somewhat more elongate; males with
   massive, beady leg segments, without adanal or accessory
   plates but with preanal plate which is continued backward
   on either side of anus as two prongs ....*Margaropus*
   (Two species; Africa; hosts, horses and giraffe)

7. Basis capituli rectangular dorsally (Figures 19.1, 19.3);
   ventral plates absent in both sexes; ornate or inornate 8
   ● Basis capituli hexagonal dorsally (Figure 19.9); ventral plates
   present or absent; usually inornate..9

8. Usually ornate; dentition of hypostome 3/3 (Figure 19.10) ....*Dermacentor* 12
   ● Inornate; dentition of hypostome 4/4 ....*Anocentor*
   (*A. nitens;* horses; Florida, Georgia, Texas to Brazil)

9. Coxae IV much larger than other coxae (see Figure 19.2);
   male without ventral plates or shield (Africa) ....*Rhipicentor*
   ● Coxae IV normal; male with ventral plates (Figure 19.11) and
   possibly showing a caudal protrusion ....*Rhipicephalus*
   (*R. sanguineus,* the brown dog tick, the only species of the
   genus known from N. America)

10. Palpi long, with 2nd segment about twice as long as broad
    (Figure 19.4); male without adanal or subanal shields...............................11
    ● Palpi long, with joints subequal; male with adanal and subanal
    shields; many species in Africa, Asia, and palearctic region ....*Hyalomma*

11. Without eyes; from reptiles; tropics and subtropics ....*Aponomma*
    ● With eyes (see Figures 19.1, 19.2); from cold- and
    warmblooded vertebrates (numerous important pests and
    vectors of disease agents of man and livestock) ....*Amblyomma*

Dermacentor of United States and Canada

12. Spurs on coxa I with proximal edges widely divergent (Figure
    19.12) 13
    ● Spurs on coxa I with facing edges parallel or nearly so (Figure
    19.2) 14

13. Scutum: male with punctations a mixture of large and small;
    female with punctations mostly large. Cervical grooves: in
    male, 3 or more times longer than broad, deep anteriorly, open
    posteriorly; in female long, deep, and expanded posteriorly.
    Adults on rabbits; immatures on rabbits and rodents ....*parumapertus*

- Scutum: male with punctations small and of uniform size; female with few enlarged punctations at periphery but remainder small. Cervical grooves short; pitlike in male; shallow in female. From peccary in southern Texas .... *halli*

14. Spiracular plate lacking dorsal prolongation (Figure 19.14) or with broad, truncate prolongation; goblet cells very large, of uniform size and moderate (75 to 85) in number. Commonly known as the winter tick; a 1-host tick of horses, cattle, deer, elk, and moose; widespread in U.S. and Canada .... *albipictus*
- Spiracular plate oval with distinct dorsal prolongation (Figure 19.15); with goblet cells minute to large and few (50 or less) to many (more than 100 in number) 15

15. Spiracular plate (Figure 19.16) with very numerous minute goblet cells, granular in appearance and subtended by wide band of fine perforations inside frame and within dorsal prolongation. American dog tick. Adults common on dogs, man, cattle, horses, and large wildlife. Eastern, central and western U.S. and central and eastern Canada .... *variabilis*
- Spiracular plate (Figure 19.15) with medium- to large-size goblet cells, with or without a narrow band of fine perforations between the cells and the frame and with a few to many fine perforations within dorsal prolongation 16

16. Male with very narrow dorsal prolongation of spiracular plate; goblet cells few (less than 50) and large; without smaller perforations between goblet cells and frame, but few perforations within dorsal prolongation; female spiracular plate similar to that of male but with slightly broader dorsal prolongation (Figure 19.15). From desert bighorn sheep in southern California, Arizona, Baja California .... *hunteri*
- Spiracular plate with moderately broad to slender dorsal (posterolateral) prolongation and with goblet cells distinct but moderate in size and number (50 to more than 100), with narrow band of minute punctations inside frame and filling dorsal prolongation (Figure 19.5) 17

17. Scutal punctations relatively uniform in size with broad areola of reddish base color subtending each punctation and producing a reddish "measled" appearance on gray patterned areas of scutum as well as on dorsal parts of leg and palpi; less pronounced in female than male. Cornua long (Figure 19.13). Pacific coast tick. California, Oregon, northern Baja California; adults on large mammals including man .... *occidentalis*
- Scutal punctations consisting of a moderate number of very large deep punctations and more numerous small punctations producing only a faint stippling of reddish dots on patterned areas. Cornua short (Figure 19.3). Rocky Mountain wood tick. Western U.S. and Canada. Adults on man, cattle, dogs, sheep, horses, and large wildlife .... *andersoni*

Argasidae

18. Margin of body more or less rounded, not thin and acute (Figures 19.18, 19.19) 19

- Margin of body thin, sharp-edged in unengorged nymphs and adults; with a sutural line separating dorsal and ventral surfaces . . . . *Argas* 20

19. Adults with integument granular, hypostome vestigial; nymphs with integument spinose (Figure 19.18), hypostome well-developed . . . . *Otobius*
   (*O. megnini* nymphs common in ears of cattle; *O. lagophilus* in ears of rabbits in western N. America)
- Adults and nymphs with integument alike, mammillated or tuberculated, and lacking spines; hypostome not vestigial . . . . *Ornithodoros* 21

Argas

20. Margins of body striate (Figure 19.21), not marked off by quadrangular "cells"; postpalpal setae absent . . . . subgenus *Argas*
   (Includes Old World pigeon tick *A. reflexus* and New World *A. brevipes* from owls, woodpeckers, and kestrel and *A. cooleyi* from cliff swallows and condor)
- Margins of body with quadrangular cells (Figure 19.22); postpalpal setae present (Figure 19.23) . . . . subgenus *Persicargas*
   (In the U.S. includes *A. persicus, A. sanchezi,* and *A. radiatus* as poultry pests. *A. giganteus* is known in the U.S. only from the larvae on wild birds. *A. miniatus* is a poultry pest in Central and S. America. Numerous other species of this widespread group occur on a variety of birds)

Ornithodoros

21. Eyes present 22
- Eyes absent 23

22. Body broad and rounded in front (see Figure 19.19); coxae I and II contiguous. Commonly feeding on camels. Attacks other animals including man. Distribution North, East, and South Africa, the Near East, India, and Sri Lanka. The "eyed tampan" . . . . *savignyi*
- Body subconical in front; coxae I and II not contiguous. Commonly feeding on cattle and deer in California and Mexico. Attacks other animals including man, to whom the tick may be venomous . . . . *coriaceus*

23. Body subconical in front 24
- Body broad and rounded in front (Figure 19.19). Attacking man, warthogs, and pigs commonly in Africa. The "eyeless tampan" . . . . *moubata*

24. Integument without obvious rounded discs; cheeks absent; dorsal humps on legs present or absent 25
- Integument with obvious rounded discs; cheeks present, dorsal humps on legs absent. Hosts rodents, but will feed on man. Central America to southern U.S. . . . . *talaje*

25. Tarsi without dorsal humps but with mild subapical dorsal protuberances. Western U.S. and Canada. Wild rodents. Bites man . . . . *hermsi*

- Tarsus I with dorsal humps and subapical dorsal protuberance. Tarsus IV without dorsal humps; subapical dorsal protuberance small or absent (Figure 19.20). Associated with rodents and lagomorphs but will feed on other hosts including man. U.S. and Mexico . . . . *parkeri; turicata*

References

Arthur, D. R. 1960. *Ticks. A Monograph of the Ixodoidea,* part V. Cambridge University Press. 251 pp.

Brinton, E. P., and D. E. Beck. 1963. *Hard Bodied Ticks of the Western United States.* Brigham Young Univ. Sci. Bull. Biol. Series, vol. 2, no. 3, parts 1–3. Brigham Young University, Provo, Utah.

Brinton, E. P., D. E. Beck, and D. M. Allred. 1965. *Identification of the Adults, Nymphs and Larvae of Ticks of the Genus* Dermacentor *Koch (Ixodidae) in the Western United States,* Brigham Young Univ. Sci. Bull. Biol. Series, vol. 5, no. 4. Brigham Young University, Provo, Utah.

Cooley, R. A. 1938. *The Genera* Dermacentor *and* Otocentor *(Ixodidae) in the United States, with Studies in Variation.* U.S. Pub. Health Serv., Natl. Inst. Health Bull. 171, Washington, D.C. 89 pp.

– 1946. *The Genera* Boophilus, Rhipicephalus *and* Haemaphysalis *(Ixodidae) of the New World.* U.S. Pub. Health Serv., Natl. Inst. Health Bull. 187, Washington, D.C. 54 pp.

Cooley, R. A., and G. M. Kohls. 1944. The Genus *Amblyomma* (Ixodidae) in the United States. *J. Parasit. 30*(2):77–111.

– 1944. *The Argasidae of North America, Central America and Cuba.* American Midland Naturalist Monograph No. 1. University of Notre Dame Press, Notre Dame, Ind. 152 pp.

– 1945. *The Genus* Ixodes *in North America.* U.S. Pub. Health Serv., Natl. Inst. Health Bull. 184, Washington, D.C. 246 pp.

Evans, G. O., J. G. Sheals, and D. MacFarlane. 1961. *The Terrestrial Acari of the British Isles.* British Museum of Natural History, London. 219 pp.

Gregson, J. D. 1956. *The Ixodoidea of Canada.* Science Service, Entomol. Div., Canada Dept. of Agriculture, Publ. 930. 92 pp.

Hoogstraal, H., 1956. *African Ixodoidea,* vol. 1, *Ticks of the Sudan.* Dept. of the Navy (U.S.), Bureau of Medicine and Surgery. U.S. Govt. Printing Office, Washington, D.C. 1101 pp.

– 1970–81. *Bibliography of Ticks and Tick-borne Diseases from Homer (about 800 B.C.) to 31 December 1969–1979,* 6 vols. Special Publ. U.S. Naval Medical Res. Unit No. 3, Cairo, Egypt.

Kaiser, M. N., H. Hoogstraal, and G. M. Kohls. 1964. The subgenus *Persicargus,* new subgenus (Ixodoidea, Argasidae, *Argas*). *A. (P.) arboreus,* new species, an Egyptian persicus-like parasite of wild birds, with a redefinition of the subgenus *Argas. Ann. Entomol. Soc. Am.* 57:60–9.

Keirans, J. E., and C. M. Clifford. 1978. The genus *Ixodes* in the United States: a scanning electron microsope study and key to the adults. *J. Med. Entomol.,* suppl. 2. 149 pp.

Kohls, G. M., H. Hoogstraal, C. M. Clifford, and M. N. Kaiser. 1970. The subgenus *Persicargas* Ixodoidea, Argasidae, *Argas*). 9 Redescription and New World records of *Argas* (P.) *persicus* (Oken), and resurrection, redescription, and records of *A. (P.) radiatus* Railliet, *A. (P.) sanchezi* Duges, and *A. (P.) miniatus* Koch, New World ticks misidentified as *A. (P.) persicus. Ann. Entomol. Soc. Am.* 63(2):590–606.

Krantz, G. W. 1978. *A Manual of Acarology,* 2nd ed. Oregon State University Book Stores, Corvallis. 509 pp.

Nuttall, G. H. F., C. Warburton, W. F. Cooper, and L. E. Robinson. 1908–26. *Ticks. A Monograph of the Ixodoidea.* parts 1–4. Cambridge University Press.

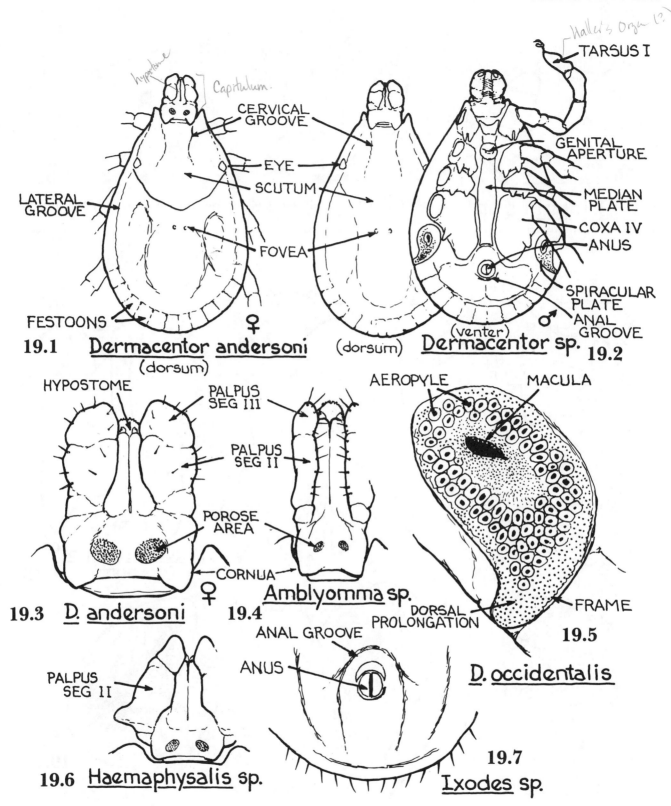

Haller's Organ (?)

TARSUS I

hypostome

Capitulum.

CERVICAL GROOVE

EYE

SCUTUM

LATERAL GROOVE

FOVEA

GENITAL APERTURE

MEDIAN PLATE

COXA IV

ANUS

SPIRACULAR PLATE

ANAL GROOVE

FESTOONS

♀

19.1   Dermacentor andersoni
(dorsum)

(dorsum)

(venter)

♂

Dermacentor sp.   19.2

HYPOSTOME

PALPUS SEG III

PALPUS SEG II

POROSE AREA

CORNUA

♀

19.3   D. andersoni

19.4   Amblyomma sp.

AEROPYLE

MACULA

DORSAL PROLONGATION

FRAME

19.5   D. occidentalis

PALPUS SEG II

ANAL GROOVE

ANUS

19.6   Haemaphysalis sp.

19.7   Ixodes sp.

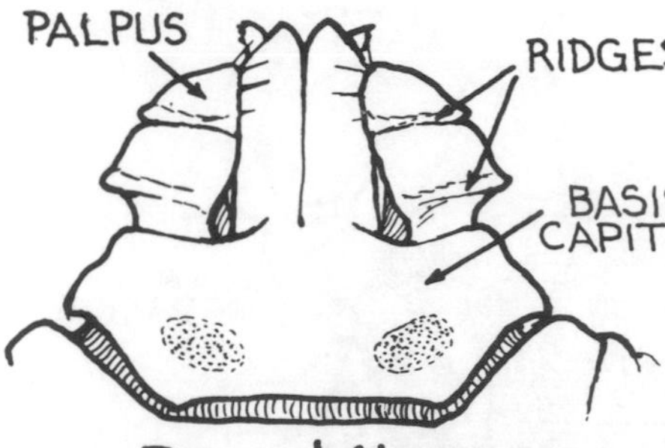

PALPUS   RIDGES

BASIS
CAPITULI

Boophilus sp.
19.8   (dorsum)

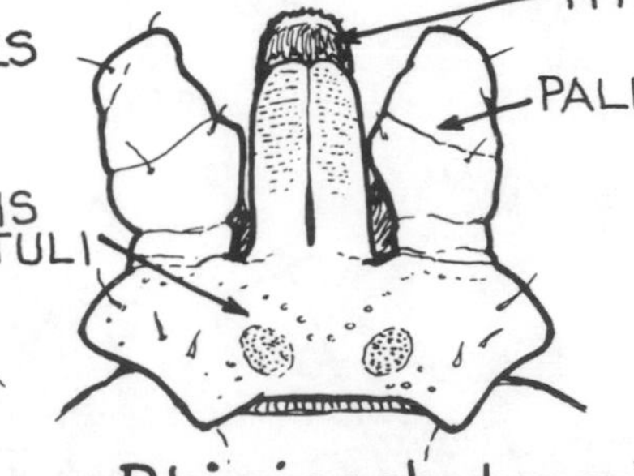

HYPOSTOME

PALPUS

Rhipicephalus sp.
19.9   (dorsum)

3/3
DENTITION

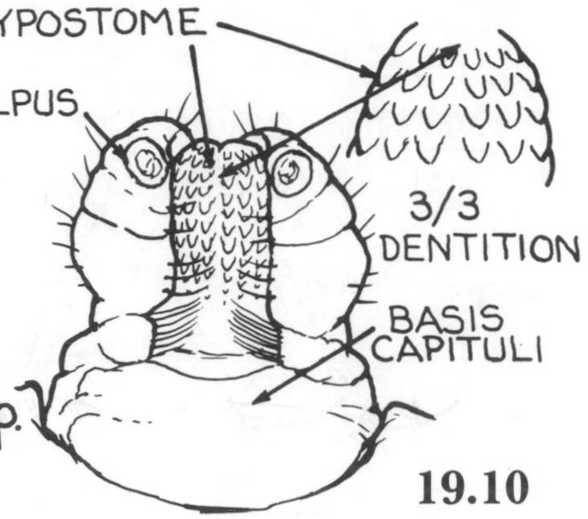

BASIS
CAPITULI

Dermacentor sp.
(venter)   19.10

COXA III
COXA IV

VENTRAL
PLATE
ANUS

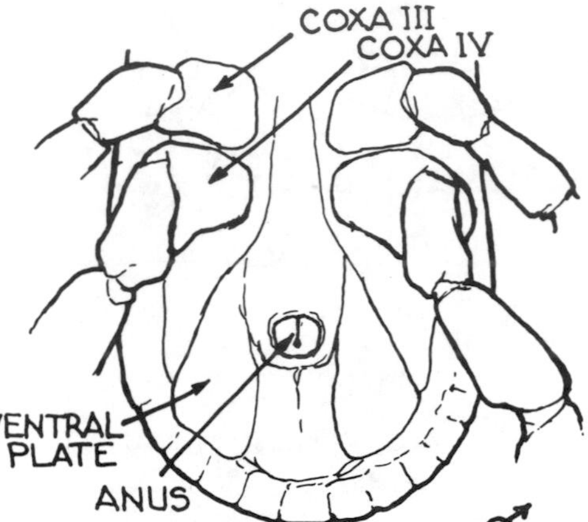

Rhipicephalus sp. ♂
19.11

D. parumapertus 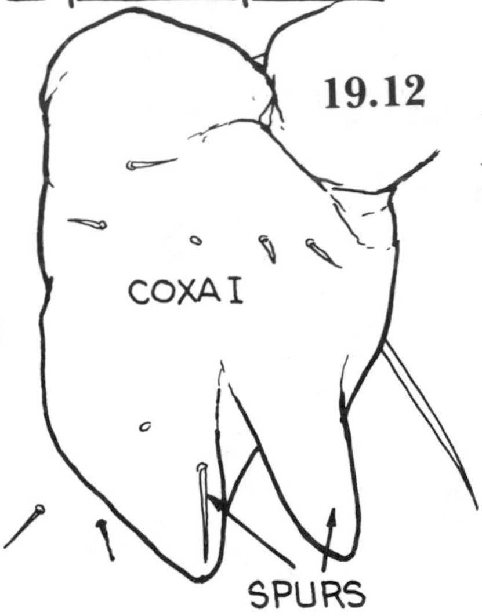
19.12

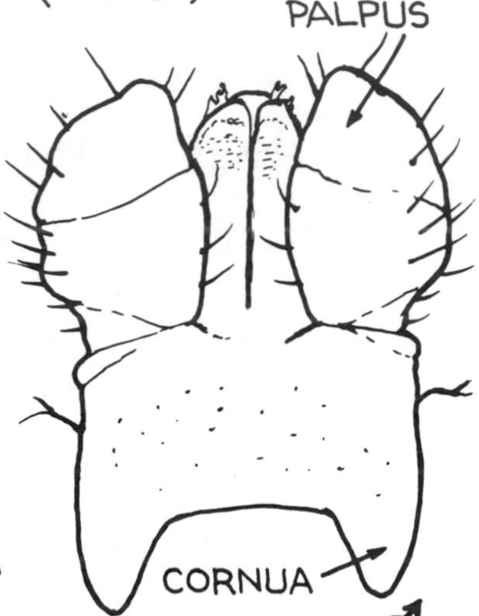

COXA I

SPURS

PALPUS

CORNUA

D. occidentalis ♂
(dorsum)   19.13

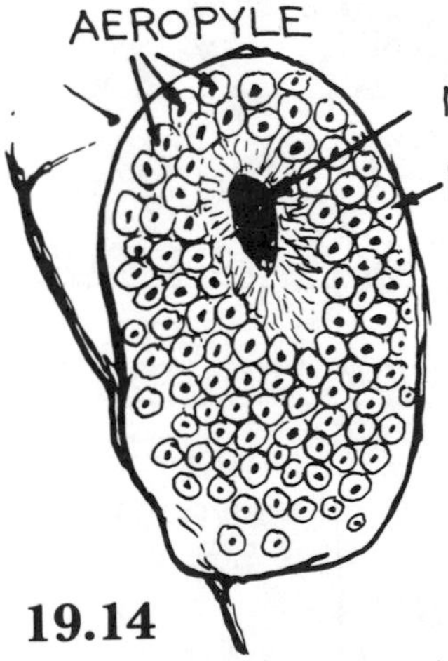

AEROPYLE

MACULA
FRAME

19.14
D. albipictus ♂

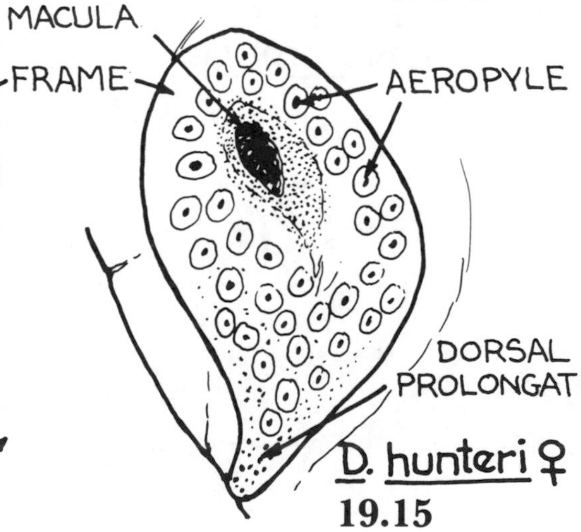

AEROPYLE

DORSAL
PROLONGATION

D. hunteri ♀
19.15

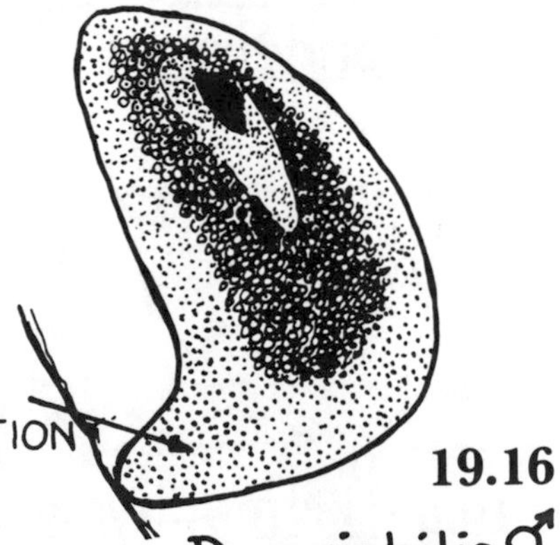

19.16

D. variabilis ♂

19.17
Dermacentor sp.
HALLER'S ORGAN — TARSUS I

CLAWS

PULVILLUS

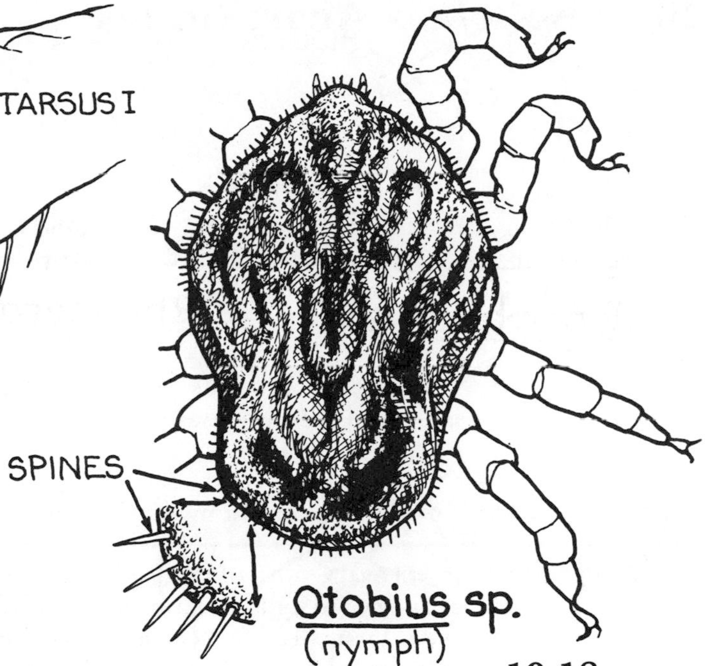

SPINES

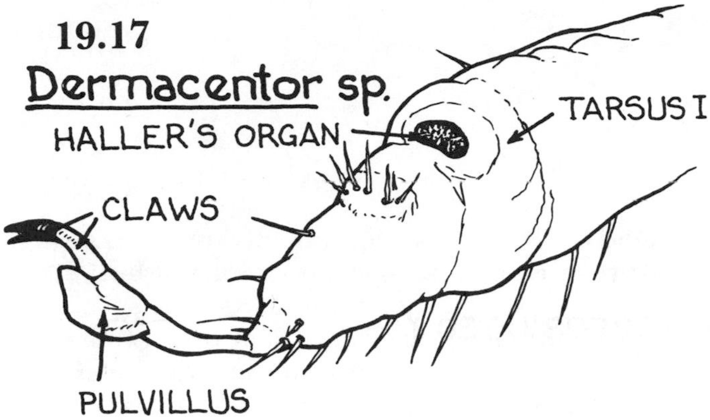

Otobius sp.
(nymph)
19.18

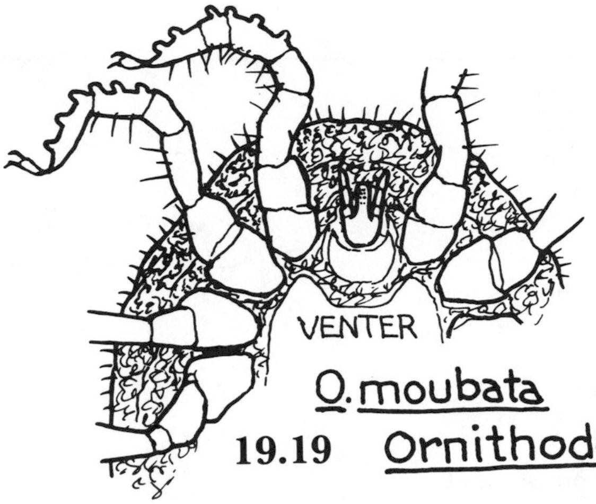

VENTER

O. moubata
19.19 ## Ornithodoros sp.

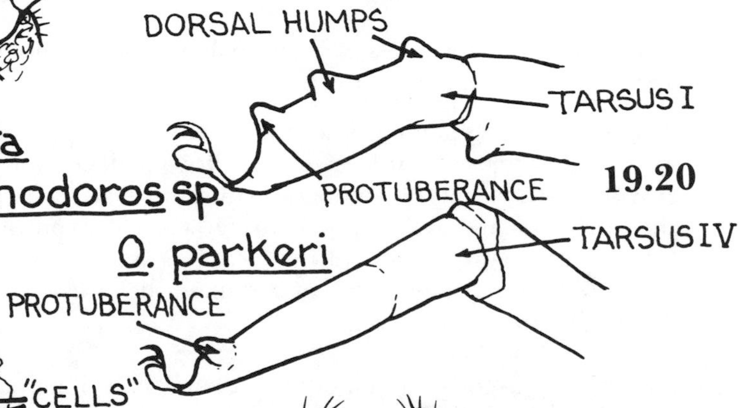

DORSAL HUMPS

TARSUS I

PROTUBERANCE
19.20

O. parkeri

TARSUS IV

PROTUBERANCE

HYPOSTOME

"CELLS"

19.21

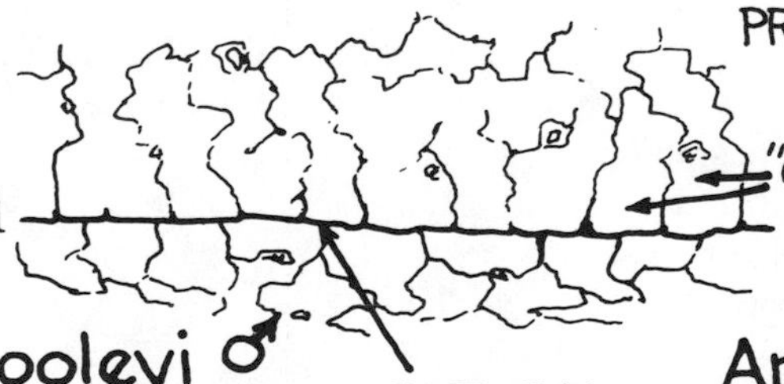

A. cooleyi ♂

SUTURAL LINE

Argas sp.

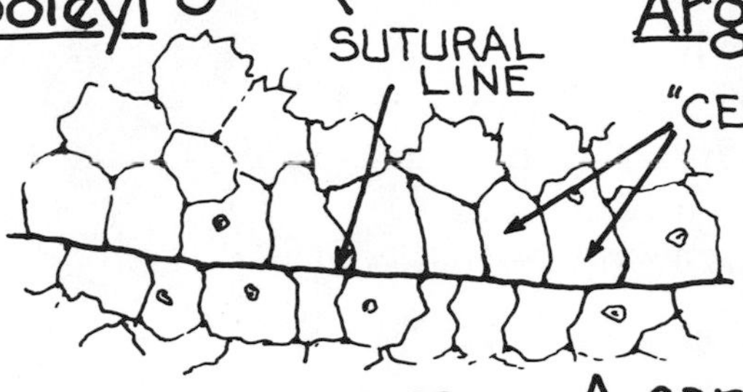

"CELLS"

A. persicus

POSTPALPAL SETA

POST-HYPOSTOMAL SETA

19.23

19.22 ## A. sanchezi ♀

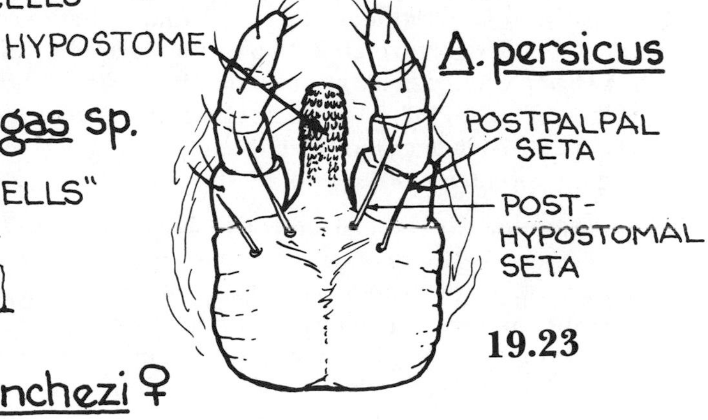

20 Subclass Acari (mites)

The higher categories of mite classification – class, subclass, order, suborder – are discussed in Exercise 19. Mites of the subclass Acari are arthropods with obscure or missing body segmentation in which the abdomen is broadly joined to the cephalothorax. They may be distinguished from ticks of the subclass Acari through the absence of Haller's organ on the tarsus of the first pair of legs, and by a hypostome not modified as a hold-fast organ and not provided with recurved teeth. In common with ticks, mites parasitic on vertebrates are characterized as adults by having 4 pairs of legs. In contrast to ticks, many mites are free-living, with a wide variety of nonparasitic food habits.

Key to certain mites of significance in medical entomology

1. With pair of stigmata at the sides of the leg-bearing region of the body, usually associated with an elongated peritreme (Figure 20.1); palpal tarsus usually bearing a tined or subdivided claw (apotele) (Figure 20.1) near the inner basal angle of tarsus . . . . .suborder Gamasida 4
   (Mesostigmata of authors)

   • Without stigmata at sides of body; palpal tarsus without apotele 2

2. Palpi small, 2-segmented; idiosoma never covered by overlapping sclerites and never vermiform; trichobothria absent on idiosoma; no stigmal openings or tracheae; chelicerae chelate for "chewing"; pretarsus of legs consisting of median claw with prominent caruncle (Figure 20.19) or an associated membranous pad, or a stalked suckerlike organ (Figure 20.10). . . . . .suborder Acaridida 15
   (Astigmata of authors)

   • Palpi usually 3- to 5-segmented and conspicuous; if palpi small and with fewer segments, then idiosoma either vermiform (i.e., *Demodex,* Figure 20.20) or with overlapping sclerites (*Pyemotes,* Figure 20.28); idiosoma often with trichobothria (Figure 20.28, 20.29); respiratory system usually present; chelicerae chelate or modified into piercing stylets or hooklike organs; pretarsus of legs not as above 3

3. Gnathosoma with conspicuous rutella (Figure 20.27); chelicerae chelate-dentate; often hard-bodied (sclerotized); with one pair of propodosomal trichobothria almost always present, composed of piliform, barbed, or clavate pseudostigmatic organs arising from conical depressions (pseudostigmata, Figure 20.29); palpi simple; tracheae, if present, never opening between chelicerae or on dorsal surface of propodosoma; free-living forms, some of which serve as intermediate hosts of tapeworms . . . . .suborder Oribatida
   (Cryptostigmata of authors)

   • Gnathosoma rarely with rutella; chelicerae rarely chelate-dentate; usually soft-bodied; if propodosomal

trichobothria present, usually without conspicuous
pseudostigmata; palpi often with tibia and tarsus forming a
"thumb-claw" (Figures 20.2, 20.4); tracheae, if present,
opening by paired stigmata between chelicerae (Figure 20.2)
or on dorsal surface of propodosoma, often with elongate
peritremes (parasitic stage of trombiculids with only 3 pairs
legs, Figure 20.21) . . . . .suborder Actinedida 34
(Prostigmata of authors)

Suborder Gamasida: superfamily Dermanyssoidea, adult females

4. Tritosternum present; genital plate present in female, not
reduced (Figure 20.1) 5
   - Tritosternum absent; genital plate lacking or greatly
   reduced 14

5. Dorsum of body with relatively few setae, arranged in definite
pattern; genitoventral plate with 8 or fewer setae (Figure 20.1)
arranged in regular pattern 6
   - Dorsum of body with dense covering of setae, without
   apparent pattern (Figure 20.3); genitoventral plate usually
   with 10 or more setae, or if less than 10 setae, they are
   irregularly arranged. Nest inhabitants or ectoparasites of birds
   and mammals . . . . .Haemogamasinae

6. Chelae strongly toothed (see Figure 20.1); corniculi well
developed (Figure 20.1); tectum a transparent flap . . . . .Laelapidae – Laelapinae 7
   - Chelae without teeth, or with weak, transparent teeth (Figures
   20.5, 20.6); corniculi poorly developed 9

7. Genitoventral plate bearing only 1 pair of setae . . . .*Androlaelaps* and related genera
   - Genitoventral plate with 4 pairs of setae (Figure 20.1) 8

8. Sternal plate longer than wide; genitoventral plate incurved
posteriorly to fit the adjacent margin of the anal plate. A
common parasite of commensal rats . . . .*Laelaps (Echinolaelaps) echidninus*
   - Sternal plate wider than long (Figure 20.1); genitoventral plate
   not as above. Many species occurring on rodents, marsupials . . . .*Laelaps* spp.

9. Sternal plate about twice as long as wide (length measured on
midline, width at level of 2nd pair of setae) . . . . .Raillietidae
Parasitic in ears of cattle . . . .*Raillietia auris*
   - Sternal plate wider than long (Figure 20.9) 10

10. Chelicerae of female very long and needlelike without obvious,
shearlike chelae (Figure 20.5) . . . . .Dermanyssidae 11
    - Chelicerae of female not needlelike, chelae shearlike (Figure
    20.6) . . . . .Macronyssidae 12

11. Dorsal plate of female entire; anal plate of female not
egg-shaped and with anal opening in its posterior end. A
common fowl mite . . . .*Dermanyssus gallinae*
    - Dorsal plate of female divided; anal plate of female
    egg-shaped and with its anal opening situated centrally.
    Ectoparasitic on mice, occasionally man . . . .*Liponyssoides sanguineus*

12. Sternal plate of female with only 2 pairs of setae (Figure 20.9).
On birds and in their nests; northern fowl mite . . . .*Ornithonyssus sylviarum*
    - Sternal plate of female with 3 pairs of setae 13

13. Dorsal plate of female narrowly attenuate posteriorly with some setae longer than width of plate (Figure 20.7). On rats, man; tropical rat mite . . . .*Ornithonyssus bacoti*
   - Dorsal plate of female not as above, but gradually decreasing in width posteriorly (Figure 20.8). On birds; tropical fowl mite . . . . *Ornithonyssus bursa*

14. Genital plate present, but reduced. Sternal shield absent or reduced. Parasitic in respiratory tracts of birds . . . . .Rhinonyssidae
   - Genital plate lacking or rudimentary (distinct in 5 known species of *Zumptiella* of rodents). Sternal shield present, elongate in female. Parasitic in respiratory tracts of mammals . . . . .Halarachnidae
     In nasal cavity of dog . . . .*Pneumonyssoides caninum*

Suborder Acaridida (= Astigmata)

15. Adults free-living; with two pairs of well-developed genital discs (acetabula) (Figure 20.18); tarsal caruncle with articulated empodial claw (Figure 20.19). Deutonymph may be heteromorphic (hypopus) occurring on or in vertebrates or invertebrates or free-living 16
   - Parasitic; genital discs greatly reduced or absent; caruncles, if present, expanded and suckerlike, without empodial claws (Figure 20.10); if caruncles absent, clawlike setae present. Without a heteromorphic hypopal stage 17

16. Tarsal claw attached by paired sclerites surrounded by pretarsus (Figure 20.19); dorsal transverse groove dividing propodosoma from hysterosoma (Figure 20.14); anal and tarsal suckers present in male. Many species and genera infest stored foods, hides. Includes forms recorded as cause of human intestinal and urogenital acariasis and contact dermatitis (*Tyrophagus, Acarus*) . . . . .Acaridae
   - Tarsal claw usually inserted at distal end of a fleshy pretarsus and joined to end of the tarsus by a single tendon (Figure 20.17); without a dorsal transverse dividing groove; anal and tarsal suckers usually absent in male. Includes many species infesting stored foods and forms that cause contact dermatitis (*Glycyphagus domesticus*) and serve as intermediate hosts of a rodent tapeworm . . . . .Glycyphagidae

17. Body usually rounded or saclike, rarely elongate; female genital opening a transverse slit lacking epigynium (genital apodemes) (Figure 20.13); genital suckers absent. Tarsi of legs I and II with pedicillate (stalked) ambulacral discs or spines (Figure 20.13). On or in skin of vertebrates . . . . .Sarcoptoidea 18
   - Body usually elongate or ovate; female genital opening occasionally a transverse slit but variable; genital suckers may be present but reduced; epigynium (genital apodemes) usually present. Legs I and II may be modified for grasping 21

18. Leg IV reduced to a small papilla-like structure, or absent. Skin parasites of bats . . . . .Teinocoptidae
   - Leg IV present, generally telescoped but well developed as in other legs 19

19. Tarsi with one or two well-developed, recurved apical prolongations ("claws"); tarsal setae always simple. On chickens and other birds . . . . .Knemidokoptidae
Four genera including
Knemidokoptes

   • Tarsi never with apical "claws"; tarsi with some setae modified as stout spines (Figures 20.13, 20.15) . . . . .Sarcoptidae 20

20. Dorsal striae broken by strong, spinelike serrations; posterior body setae short, stout, lanceolate; anus terminal. Causes scabies or itch in man, mange in domestic animals . . . .*Sarcoptes scabiei*
   • Dorsal striae not broken by spinelike serrations; posterior dorsal body sctac not as above; anus dorsal. On cats and rabbits . . . .*Notoedres cati*

21. Neither legs nor mouthparts adapted for grasping or clasping 22
   • With some legs and/or mouthparts adapted for grasping tissues or hairs of mammalian hosts . . . . .Listrophoroidea 33

22. Female genital opening a longitudinal slit or narrow inverted V; lacking genital suckers . . . . .Cytoditoidea 23
   • Female genital opening usually an inverted U, Y, or V shape, or if opening is longitudinal, there is a strong epigynium bordering slit anteriorly and laterally; genital suckers present but reduced 24

23. Opisthosoma very short; legs IV arising on posterior fourth of body. Small, oval, nude mites in tissue, particularly lining of air sacs of birds including poultry . . . . .Cytodytidae
Cytodites nudus

   • Opisthosoma not foreshortened, often elongate; legs IV arising anterior to posterior fourth of body. In subcutaneous tissues of birds including poultry . . . . .Laminosioptidae
Laminosioptes cysticola

24. Parasites of mammals, occasionally free-living or nidicolous . . . . .Psoroptoidea 25
   • Parasites or paraphages of birds. Feather and skin mites; hyperparasites of hippoboscids, mallophagans 28

25. Legs III and IV of adults arising ventrally or laterally; legs III usually with long terminal setae (Figure 20.10); male with anal suckers (Figure 20.11). Skin parasites of mammals . . . . .Psoroptidae 26
   • Legs III and IV of female arising ventrally and without long terminal setae; male with or without anal suckers. House dust mites, living also in stored food products, on birds and mammals or in their nests . . . . .Pyroglyphidae
Dermatophagoides spp.

26. Pretarsi with short stalks; female genital opening an almost transverse slit 27
   • Pretarsi with long stalks on legs I, II, and IV in female (Figure 20.10) and on legs I, II, and III of male; female genital opening an inverted U; on cattle, sheep, goats, horses, and ears of rabbits. Sweatman (1958) lists 5 species . . . .*Psoroptes*

27. All tarsi of male and all but tarsus III of female with stalked pretarsi; coxal apodemes I and II free; on domestic animals . . . .*Chorioptes bovis*

- All tarsi of male and tarsi I and II of female with short, stalked pretarsi; legs IV much reduced; coxal apodemes I and II fused ventrally; in ears of dogs and cats . . . .*Otodectes cynotis*

28. Legs attached submedially; coxae of legs III and IV not visible in dorsal view (Figure 20.16); coxal apodemes usually fused to form a rounded network, the coxal field; tibiae and tarsi of anterior legs with keels (Figure 20.16); ambulacra large, leaflike, on short, thick pedicles; setae of posterior body margin usually widened and leaflike. Feather mites . . . . .Freyanidae
- Legs attached near lateral margins of body, coxae of legs III and IV visible from above (Figure 20.12); coxal apodemes developed variously, but the elements rarely fused, and, if so, the coxal field small and not rounded 29

29. Posterior of females and most males divided by a deep cleft; posterior lobes armed with long (often modified) setae, and may have hyaline, swordlike or leaflike appendages (Figure 20.12). Feather mites . . . . .Proctophyllodidae
- Posterior of female rounded, without pronounced lobes; posterior of male body variously shaped, but never bearing swordlike or leaflike appendages 30

30. Transparent leaflike projections, recurved hooks or strong, sharply pointed setae on the ventroexternal surface of the 3rd and 4th segments of anterior legs. Central setae of ventral surface of opisthosoma long and strong, but hyaline. Feather mites . . . . .Analgidae (= Analgesidae)
  (Includes *Megninia* species as feather mites of poultry)
- Without ventroexternal excrescenses or projections on anterior legs. Central setae of ventral surface of opisthosoma very small 31

31. Anterior vertical setae usually present; cuticula strongly sclerotized; body shields usually well developed; males with adanal discs; living on or in feathers of birds 32
- Anterior vertical setae absent; cuticula membranous, finely striated; body shields poorly developed; males with or without adanal discs; living on birds including poultry and as hyperparasites on hippoboscids, mallophagans . . . . .Epidermoptidae
  Epidermoptes bilobatus

32. Female with genital apodemes reduced or vestigial; pregenital sclerite (epigyneum) lacking or very reduced. Copulatory suckers of male often reduced and rarely absent . . . . .Dermoglyphidae
- Female with genital apodemes well sclerotized; epigyneum well developed and sclerotized forming a more or less regular arc around 0.25 to 0.75 of genital circumference. Copulatory suckers always present in male . . . . .Pterolichidae

33. Palpi modified for clasping hair. Legs III and/or IV not modified for hair clasping. Parasites of rodents, lagomorphs, carnivores, and insectivores . . . . .Listrophoridae
- Gnathosoma not modified for clasping hair. Legs III and/or IV modified for hair clasping. Fur mites of rodents . . . . .Myocoptidae

Suborder Actinedida (=Prostigmata)

34. Body elongate, vermiform, annulate, without setae (Figure
    20.20); in skin pores of mammals . . . . .Demodicidae – *Demodex* spp.
    - Body not as above 35

35. With 4 pairs of legs 43
    - With 3 pairs of legs (Figure 20.21); usually attached to host;
    larval forms, most of which feed on terrestrial vertebrates . . . . .Trombiculidae 36

36. Coxa I with single seta; all legs with 7 segments (Figure
    20.21) 37
    - Coxa I with two setae; all legs with 6 segments; scutum with
    anteromedian projection (nasus) and paired submedian setae
    (Figure 20.22). Several species, some of which attack man . . . .*Odontacarus, Acomatacarus*
    subfamily Leeuwenhoekiinae

37. Scutum (Figure 20.23) with 1 anteromedian seta but without
    an anteromedian projection (nasus); posterolateral setae
    usually on scutum . . . . .subfamily Trombiculinae (in part) 38
    - Scutum with 2 anteromedian setae and a nasus; eyes
    prominent. On domestic chicken, Brazil . . . .*Apolonia tigipoensis*
    subfamily Apoloniinae

38. Sensillae of scutum inflated (Figure 20.24) 41
    - Sensillae flagelliform (Figure 20.23) 39

39. Mastitarsala III (a long nude seta on dorsal side of tarsus of leg
    III) present 40
    - Mastitarsala III absent. On birds and mammals, including man.
    Includes vectors of typhus . . . .*Leptotrombidium*
    L. akamushi

40. Palpal claw three-pronged; scutum pentagonal. Parasitic on
    small mammals . . . .*Neotrombicula*
    - Palpal claw bifurcate (see Figure 20.4); scutum subquadrate
    (see Figure 20.23). On many vertebrates, including man . . . . *Eutrombicula*

41. Cheliceral blade with a single subapical dorsal tooth (Figure
    20.26) 42
    - Cheliceral blade with row of dorsal teeth (Figure 20.4).
    Common on rodents, occasionally on birds or man . . . .*Schoengastia*

42. Scutum partially covered by cuticular striations; sensilla usually
    globose. Common parasites of birds . . . .*Neoschoengastia*
    - Scutum not covered by striations; sensilla not usually globose.
    Primarily parasites of rodents and birds in southeast Asia and
    America, occasionally attack man . . . .*Euschoengastia*

43. First pair of legs highly modified for clasping hairs of hosts
    (Figure 20.25); fur mites of rodents, insectivores, and bats . . . . .Myobiidae
    - First pair of legs normal, for walking 44

44. Gnathosoma conspicuous; stigma opening at base of chelicerae
    (Figure 20.2) 45
    - Gnathosoma not conspicuous; chelae tiny, stylet-like; stigma of
    female opening on propodosoma, behind gnathosoma; gravid
    female enormously distended and easily visible to naked eye;
    parasite of insect larvae; bites man (Figure 20.28) .Pyemotidae – *Pyemotes tritici*

45. Palpi with thumb-claw complex (Figure 20.2); tarsi with a double row of tenent hairs (Figure 20.31) ....Cheyletiellidae, Cheyletidae 46
   - Palpi simple; stubby, radiating legs each with a strong, ventral, femoral sclerotic hook; tarsi each with two claws and lacking tenent hairs ....Psorergatidae
     Skin parasite of sheep ....*Psorergates ovis*

46. Sensory organ on genu I not heart-shaped or depressed at apex 47
   - Sensory organ on genu I heart-shaped and depressed at apex (Figure 20.30). Skin parasite on dogs ....*Cheyletiella yasguri*

47. Posterior margin of hysterosoma invaginated; sensory organ on genu I conical; skin parasite of cats ....*Cheyletiella blakei*
   - Posterior margin of hysterosoma not invaginated; sensory organ on genu I not conical; on hares and rabbits ....*Cheyletiella* spp.
     (*C. parasitivorax,*
     C. strandtmanni, C. furmani)

References

Baker, E. W., J. H. Camin, F. Cunliffe, T. A. Wooley, and C. E. Yunker. 1958. *Guide to the Families of Mites.* Inst. Acarology, University of Maryland. 242 pp.

Baker, E. W., T. M. Evans, D. J. Gould, W. B. Hull, and H. L. Keegan. 1956. *A Manual of Parasitic Mites of Medical or Economic Importance.* National Pest Control Assoc., New York. 170 pp.

Baker, E. W., and G. W. Wharton. 1952. *An Introduction to Acarology.* Macmillan, New York. 465 pp.

Brennan, J. M., and M. L. Goff. 1977. Keys to the genera of chiggers of the Western Hemisphere (Acarina: Trombiculidae). *J. Parasitol. 63*:555–66.

Dubinin, W. B. 1953. *Feather Mites (Analgesoidea).* Part II, Families Epidermoptidae and Freyanidae. Acad. Sci. U.S.S.R., Moscow (in Russian). 411 pp.

Evans, G. O., J. G. Sheals, and D. MacFarlane. 1961. *The Terrestrial Acari of the British Isles.* British Museum of Natural History, London.

Fain, A. 1968. Etude de la variabilité de *Sarcoptes scabiei* avec une revision des Sarcoptidae. *Acta Zoologica et Pathologica Antverpiensia 47.* 196 pp.

Fain, A., and P. Elsen. 1967. Les acariens de la famille Knemidokoptidae producteurs de gale chez les oiseaux. *Acta Zoologica et Pathologica Antverpiensia 45.* 142 pp.

Hughes, A. M. 1961. *The Mites of Stored Food.* Her Majesty's Stationery Office, London.

Johnston, D. E. 1968. *An Atlas of Acari I The Families Parasitiformes and Opilioacariformes.* Publ. 172 Acarology Laboratory, Columbus, Ohio. 110 pp.

Krantz, G. W. 1978. *A Manual of Acarology,* 2nd ed. Oregon State University Book Stores, Corvallis. 509 pp.

Smiley, R. L. 1970. A review of the family Cheyletiellidae (Acarina). *Ann. Entomol. Soc. Am. 63*:1057–78.

Strandtmann, R. W., and G. W. Wharton. 1958. *Manual of Mesotigamatid Mites Parasitic on Vertebrates.* Inst. Acarology, University of Maryland. 330 pp., 69 plates.

Sweatman, G. K. 1957. Life history, non-specificity, and revision of the genus *Chorioptes,* a parasitic mite of herbivores. *Canad. J. Zool.* 35:641–89.

– 1958. On the life history and validity of the species in *Psoroptes,* a genus of mange mites. *Canad. J. Zool.* 36:905–29.

Vercammen-Grandjean, P. H., and R. Langston. 1976. *The Chigger Mites of the World.* vol III, Leptotrombidium *Complex.* Section A, *Leptotrombidium* sensu stricto. G. W. Hooper Foundation, University of California, San Francisco. 612 pp.

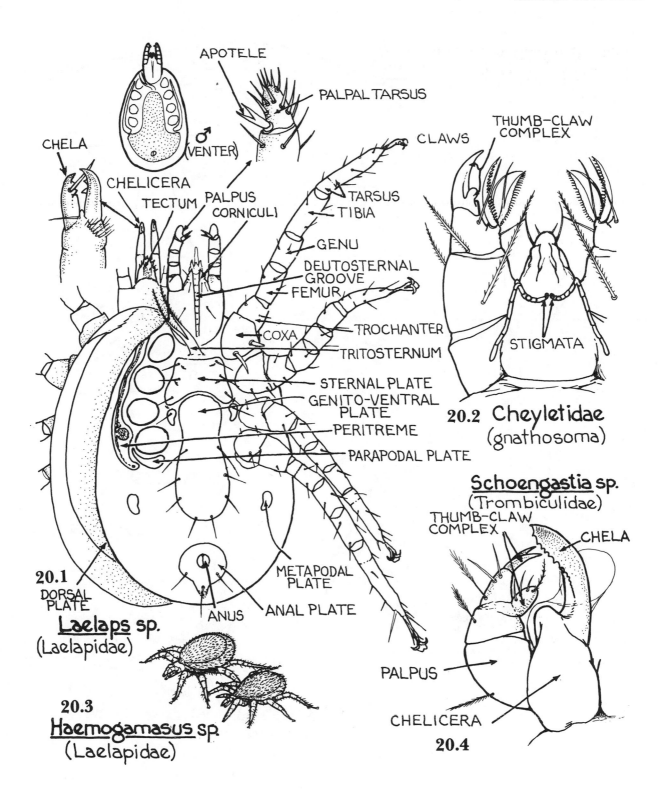

CHELA

APOTELE

PALPAL TARSUS

♂(VENTER)

CHELICERA

TECTUM

PALPUS
CORNICULI

CLAWS

THUMB-CLAW
COMPLEX

TARSUS
TIBIA

GENU

DEUTOSTERNAL
GROOVE
FEMUR

TROCHANTER

COXA

TRITOSTERNUM

STERNAL PLATE

GENITO-VENTRAL
PLATE

PERITREME

PARAPODAL PLATE

STIGMATA

20.2 Cheyletidae
(gnathosoma)

Schoengastia sp.
(Trombiculidae)

THUMB-CLAW
COMPLEX

CHELA

20.1
DORSAL
PLATE

Laelaps sp.
(Laelapidae)

METAPODAL
PLATE

ANUS   ANAL PLATE

PALPUS

CHELICERA

20.4

20.3
Haemogamasus sp.
(Laelapidae)

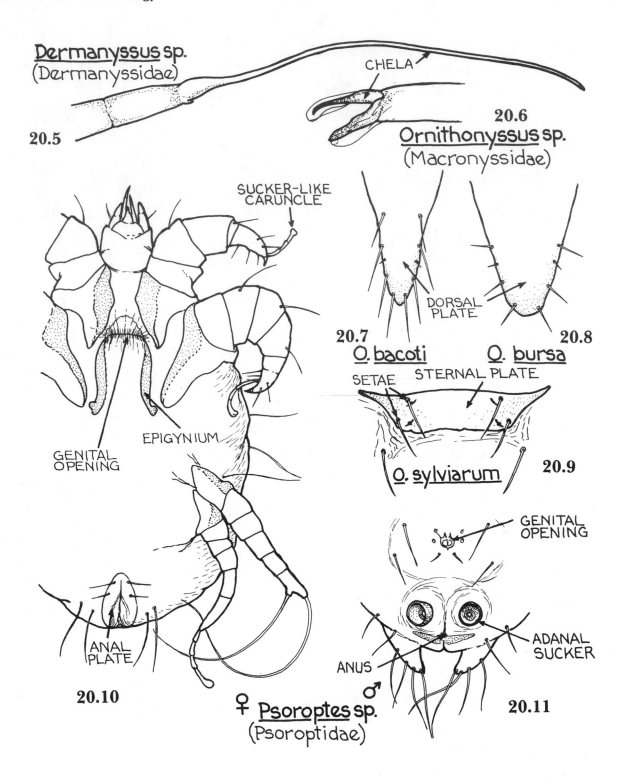

Dermanyssus sp.
(Dermanyssidae)

CHELA

20.5

20.6

Ornithonyssus sp.
(Macronyssidae)

SUCKER-LIKE
CARUNCLE

DORSAL
PLATE

20.7   20.8

O. bacoti   O. bursa

SETAE   STERNAL PLATE

EPIGYNIUM

GENITAL
OPENING

O. sylviarum   20.9

GENITAL
OPENING

ANAL
PLATE

ANUS   ADANAL
SUCKER

20.10

♀ Psoroptes sp. ♂
(Psoroptidae)   20.11

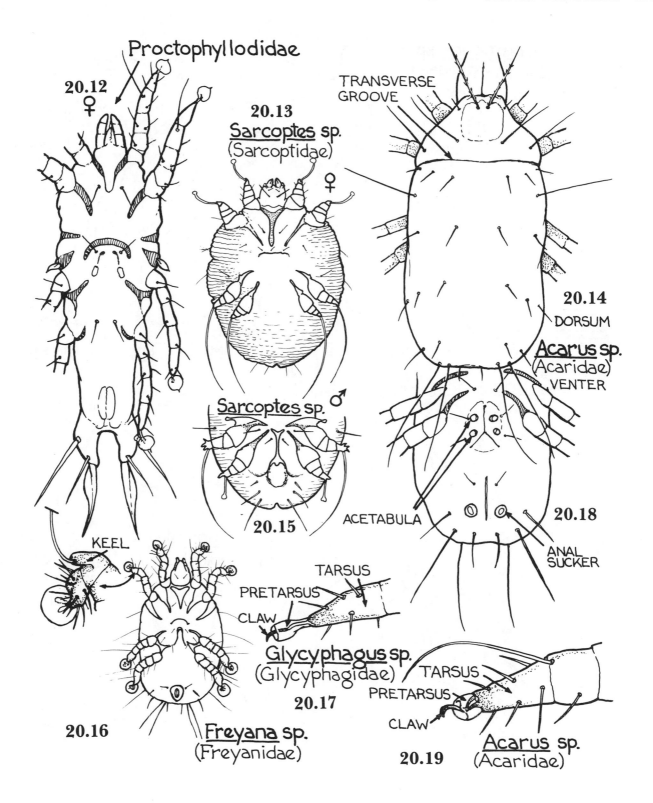

Proctophyllodidae

20.12 ♀

20.13
Sarcoptes sp.
(Sarcoptidae) ♀

Sarcoptes sp. ♂

20.15

KEEL

20.16

Freyana sp.
(Freyanidae)

TARSUS
PRETARSUS
CLAW

Glycyphagus sp.
(Glycyphagidae)

20.17

TRANSVERSE
GROOVE

20.14
DORSUM

Acarus sp.
(Acaridae)
VENTER

ACETABULA

20.18

ANAL
SUCKER

TARSUS
PRETARSUS
CLAW

Acarus sp.
(Acaridae)

20.19

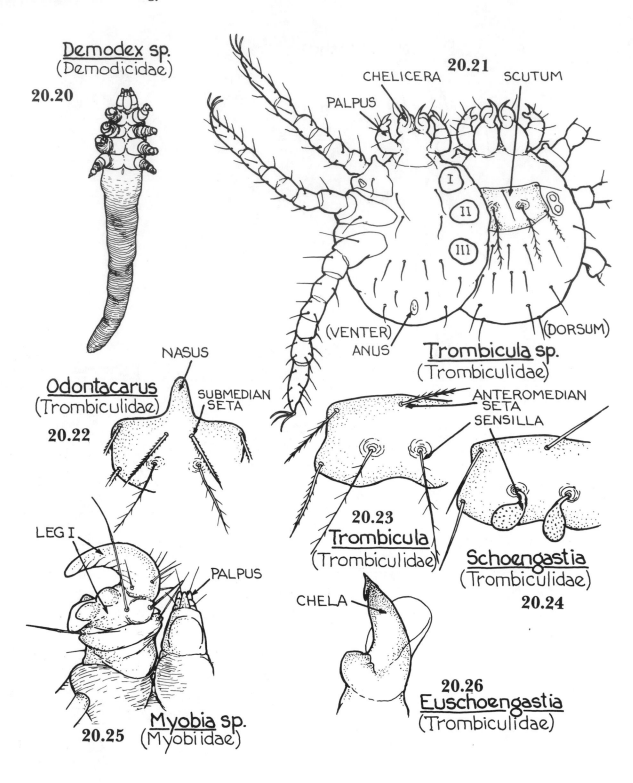

Demodex sp.
(Demodicidae)

20.20

20.21

CHELICERA   SCUTUM

PALPUS

I

II

III

(VENTER)   (DORSUM)

ANUS

Trombicula sp.
(Trombiculidae)

NASUS

SUBMEDIAN
SETA

Odontacarus
(Trombiculidae)

20.22

ANTEROMEDIAN
SETA
SENSILLA

20.23
Trombicula
(Trombiculidae)

Schoengastia
(Trombiculidae)

20.24

LEG I

PALPUS

CHELA

20.26
Euschoengastia
(Trombiculidae)

20.25   **Myobia** sp.
(Myobiidae)

PALPUS
RUTELLUM

20.27
Oribatida

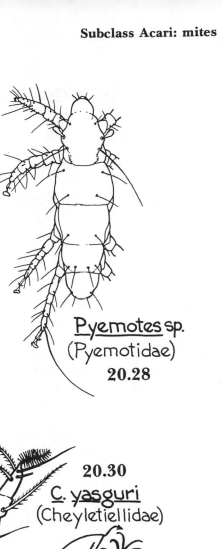

Pyemotes sp.
(Pyemotidae)
20.28

PALPUS

TRICHOBOTHRIUM

Oribatida
20.29
PSEUDOSTIGMATA

20.30
C. yasguri
(Cheyletiellidae)

GENU I

TENENT
HAIRS

TARSUS

20.31
Cheyletiella sp.
(Cheyletiellidae)

21 Venomous arthropods

Spiders, scorpions, and certain Hymenoptera (bees, wasps, hornets, and fire ants) constitute the three major groups of venomous arthropods. Keys are provided in this exercise to the important families of each of these groups and to the North American genera of vespid wasps.

Spiders

Spiders form the order Araneida, including some 30,000 known species. Like the Acari they are arachnids. They possess chelicerae and lack antennae and opposable mandibles. Their abdomen is attached to the cephalothorax by a narrow "waist," the pedicle, and shows little or no external segmentation. The cephalothorax is protected dorsally by a shieldlike carapace and has four pairs of legs and one pair of pedipalps (Figure 21.16). The tips of the male pedipalps are bulbose, complicated structures used in copulation. Only females possess the epigynum (Figure 21.16).

All spiders are predaceous and poisonous, but those dangerous to man are few in number. Spiders, with few exceptions, have eyes and are attracted by movement of their prey. Their ability to spin silk is used in various ways, including the protecting of eggs, the snaring of prey, and movement from place to place.

The following key includes only the most common spider families of North America along with a list of species whose bites are known to be dangerous to man. As indicated in Exercise 3, spiders should be stored in alcohol.

Key to common families of North American spiders of potential medical importance[a]

1. Chelicerae paraxial (fangs articulate parallel to each other) (Figure 21.1); two pairs of book lungs present . . . . suborder Orthognatha 2
   - Chelicerae diaxial (fangs oppose each other in articulation) (Figure 21.16); usually one pair of book lungs with one or two pairs of tracheal spiracles . . . . suborder Labidiognatha 4

2. Labium fused to sternum (Figure 21.1); endites of pedipalp coxae strongly developed (Figure 21.1); 6 spinnerets; thoracic furrow transverse and deep (see Figure 21.1) (purse web spiders) . . . . Atypidae
   - Labium not fused to sternum: coxal endites of pedipalps not strongly developed 3

3. Tarsi with 2 claws and claw tufts (see Figure 21.9); large (over 40 mm in total length); hairy robust appearance (tarantula spiders) . . . . Theraphosidae
   - Tarsi with 3 claws, without claw tufts; chelicerae with teeth on apical margin (trap-door spiders) . . . . Ctenizidae

4. Cribellum present (Figure 21.2); calamistrum (consisting of 2 rows of curved hairs on metatarsus IV) present (Figure 21.3); 8 eyes, all light in color . . . . Amaurobiidae
   - Cribellum and calamistrum absent 5

5. Tarsus IV with a ventral "comb" composed of a row of curved serrated bristles (Figure 21.4); abdomen often globose (Figure 21.7) (comb-footed spiders, includes the widow spiders) . . . . Theridiidae
   - No "comb" on tarsus IV even though abdomen may be globose 6

6. With 6 eyes 7
- With 8 eyes 8

7. Eyes in 2 triads (Figure 21.6); metatarsi and tarsi long, slender, and flexible (basement spiders) ....Pholcidae (in part)
- Eyes in 3 diads, median diad in advance of lateral diads (Figure 21.5); thoracic furrow conspicuous and longitutinal on flat carapace (brown spiders) ....Loxoscelidae

8. Lateral eyes triad, median eyes paired (Figure 21.6); metatarsi and tarsi long, slender, and flexible (basement spiders) ....Pholcidae (in part)
- Eight eyes not arranged as above...9

9. Tarsi with 2 claws; claw tufts (Figure 21.9) may or may not be present 10
- Tarsi with 3 claws (Figure 21.8), without claw tufts 13

10. Eyes in 3 rows, anterior row consisting of 4 eyes with median pair the largest (Figure 21.11), 2 eyes in each other row (jumping spiders) ....Salticidae
- Eyes in 2 rows, posterior may be slightly recurved (Figure 21.10) 11

11. Legs I and II laterigrade (turned 90° at attachment so ventral surface becomes anterior) giving overall crablike appearance) (Figure 21.15) (crab spiders) ....Thomisidae
- Legs I and II prograde as in most spiders 12

12. Anterior spinnerets contiguous, or nearly so, and not distinctly different from posterior ones (sac spiders) ....Clubionidae
- Anterior spinnerets separated by at least the diameter of 1 of them, and differing in appearance from posterior spinnerets (Figure 21.12) ....Gnaphosidae

13. Posterior 6 eye arrangement hexagonal (Figure 21.13) (consisting of procurved posterior row and recurved anterior row); legs with numerous prominent spines (lynx spiders) ....Oxyopidae
- Eye grouping not as above 14

14. Tarsi with trichobothria (Figure 21.8); lip of labium not rebordered along front edge 15
- Tarsi without trichobothria; anterior edge of labium rebordered (thickened) (Figure 21.14) 17

15. With a single row of tarsal trichobothria increasing in length distally (Figure 21.8); posterior spinnerets greatly elongated (up to twice the length of anterior spinnerets); trochanters not notched (funnel web spiders) ....Agelenidae
- Tarsi with trichobothria in no characteristic arrangement or gradation in length; trochanters with distinct ventral notch (Figure 21.16) 16

16. Posterior eye row strongly recurved (forming what appears as 2 rows of 2 eyes each); median tarsal claw smooth (wolf spiders) ....Lycosidae
- Posterior eye row not strongly recurved (only slightly so); median tarsal claw toothed (nursery web spiders) ....Pisauridae

17. Clypeus as high as or higher than the median ocular area (Figure 21.17); eyes may differ in size (bowl and doily spiders) ....Linyphiidae

- Clypeus lower in height than that of median ocular area (Figure 21.18); eyes of similar size (weavers of orb webs) 18

18. Femora with trichobothria (at least 1 or 2); large powerful chelicerae; abdomen elongate (long-jawed orb weavers) . . . . .Tetragnathidae
   - Femora without trichobothria; chelicerae normal size, often with a lateral boss (Figure 21.18); abdomen oval, tapering, or ornately pointed (orb weavers) . . . . .Araneidae

---

^aPrepared by G. Uetz.

Dangerous and semidangerous spiders in the United States

Scientific name	Common name	Nature of venom
Latrodectus group (widow spiders; family Theridiidae)		
Latrodectus mactans mactans	Black widow spiders	Neurotoxic
Latrodectus mactans hesperus		
Latrodectus mactans texanus		
Latrodectus bishopi	Red widow	Neurotoxic
Latrodectus geometricus	Brown widow	Neurotoxic
Latrodectus variolus	Northern widow	Neurotoxic
Steatoda grossa	House spider	Neurotoxic
Loxosceles group (brown house spiders, recluse spiders, violin spiders; family Scytotidae)	Brown recluse	Cytotoxic
Loxosceles reclusa	South American house spider	Cytotoxic
Loxosceles rufescans		Cytotoxic
Loxosceles rufipes	Brown house spider	Cytotoxic
Loxosceles arizonica		Cytotoxic
Loxosceles devia		Cytotoxic
Wolf spiders (superfamily Lycosoidea)		
Lycosidae		
Lycosa caroliniensis		Slightly neurotoxic
Lycosa punctulata		Slightly neurotoxic
Lycosa miami		Slightly neurotoxic
Agelenidae		
Tegenaria domestica	House spider	?
Pisauridae		
Dolomedes sp		?
Chiracanthium group (family Clubionidae)		
Chiracanthium inclusum	House spider	?
Chiracanthium diversum		Cytotoxic
Miscellaneous		
Amaurobius ferox		
Amaurobius bennetti		Neurotoxic
Aphonopelma sp.	Common tarantula	?
Argiope aurantia	Golden garden spider	Neurotoxic
Pamphobeteus sp.		?
Peucitia viridans	Poison squirting spider	?
Ummidia audouinni		?

Scorpions

Scorpions also are predaceous arachnids, and their defensive, toxic stings result from accidental contact. The most dangerous of the six scorpion families is the Buthidae, some of which have powerful, neurotoxic venom. Scorpion stings are of major importance in the Neotropics as agents of envenomization.

Key to families of Scorpionida of the world[a]

1. Sternum consisting of two small, separated plates, inconspicuous (S. America, Australia) ....Bothriuridae
● Sternum large, entire, conspicuous (Figure 21.19) 2

2. Tarsus with one spur (articulating in membrane at base of tarsus) 3
● Tarsus with two spurs (Figure 21.21) 4

3. With a subterminal spine on the telson (see Figure 21.20) (Mideast, Mexico) ....Diplocentridae
● Without a subterminal spine on the telson (world tropics and subtropics) ....Scorpionidae

4. With two pairs of marginal ocelli (Figure 4.1) (N. and S. America, S. Europe, N. Africa, Asia) ....Chactidae
● With three or more pairs of marginal ocelli 5

5. Lateral margins of sternum subparallel (Figure 21.19); without a subterminal spine on the telson (Mideast, Asia, Central and S. America, U.S.) ....Vejovidae
● Sternum subtriangular (margins not subparellel); with or without a subterminal spine on the telson (Figure 21.20) (world tropics) ....Buthidae

[a] Modified from Byalynitskii-Birulya (1917).

Hymenoptera

Among the Hymenoptera, those living in social groups are the most dangerous because multiple stings often result from contact with the defended nest site. The social vespids and the fire ants are of greatest importance in North America. Their sting is formed from a modified ovipositor in worker-caste individuals.

Key to families of medically important stinging Hymenoptera, adults only[a]

1. Possessing wings 2
● Lacking wings 13

2. First abdominal segment nodelike (Figure 21.22), joining of abdomen and thorax clearly constricted (ants) ....Formicidae (in part)[b]
● First abdominal segment not nodelike, may or may not be joined to thorax as constriction 3

3. Thoracic hairs simple, unbranched; first (proximal) segment of hind tarsus not broadened or thickened (wasps) 4
● Thoracic hairs branched or plumose; first segment of hind tarsus broader and thicker than distal segments (bees) 9

4. Pronotum extended caudally to end at, or near, forewing base (tegula) and without rounded lobe posteriorly on each side (Figure 21.24) 5
   - Pronotum more truncate, collarlike, with a rounded lobe posteriorly on each side (Figure 21.25) and not extended to wing base (tegula); a large group of predaceous wasps (cicada killer, mud daubers, digger wasps) ....Sphecidae

5. First discoidal cell of forewing longer than submedian cell (Figure 21.26); forewing usually folded in repose (paper hornets, wasps, and yellowjackets) ....Vespidae 14[b]
   - First discoidal cell of forewing shorter than submedian cell (Figure 21.27); forewing rarely folded 6

6. Mesopleuron with transverse suture; legs long, hind femur extending beyond apex of abdomen (spider wasps) ....Pompilidae
   - Without this combination of characters 7

7. Bases of meso- and sometimes metacoxae covered by plates (Figure 21.30). 8
   - Bases of meso- and metacoxae exposed, not covered by plates; body densely set with long colorful hair. Males of velvet ants; most are parasitic on other Hymenoptera ....Mutillidae (in part)

8. Bases of meso- and metacoxae covered by plates (Figure 21.28); wing membrane with corrugated wrinkles distal to veins (Figure 21.29). Parasitic on scarab beetle larvae ....Scoliidae
   - Bases of mesocoxae only covered by plates (Figure 21.30); wing membrane beyond veins not corrugated. Includes many beneficial parasites of beetles ....Tiphiidae

9. Hind tibia without spurs (honeybees and near relatives) ....Apidae
   - Hind tibia with one or two spurs 10

10. Hindwing with a jugal lobe (Figure 21.31) 11
    - Hindwing lacking a jugal lobe; large hairy bees usually with contrasting black and yellow or orange pile (bumble bees) ....Bombidae[b]

11. Jugal lobe of hindwing longer than submedian cell (Figure 21.31); forewing with strongly arched basal vein (Figure 21.31) (metallic solitary bees) ....Halictidae
    - Jugal lobe of hindwing shorter than submedian cell 12

12. Forewing with two submarginal cells (Figure 21.32), pilose bees (leaf cutting bees) ....Megachilidae
    - Forewing with three submarginal cells; small- to large-size bees with or without dense hairy vesture (cuckoo, digger, and carpenter bees) ....Anthophoridae

13. First abdominal segment nodelike (Figure 21.22); without densely set body hairs (ants) ....Formicidae (in part)[b]
    - First abdominal segment not nodelike, body densely set with long colorful hairs. Most are parasitic on Hymenoptera (velvet ants) (Figure 21.23) ....Mutillidae (in part)

Genera of social Vespidae of America north of Mexico[c]

14. First abdominal segment convex in lateral view (Figure 21.33); clypeus pointed or narrowly truncate apically (frontal view) (Figure 21.34) worldwide distribution (paper wasps) (Polistinae) ....*Polistes*

- First abdominal segment vertically flattened in lateral view (Figure 21.36); clypeus with apex emarginate, broadly truncate or broadly rounded apically (frontal view) (Figure 21.37) 15

15. Vertex long, distance from lateral ocellus to occipital carina much greater than distance between lateral ocelli (Figure 21.35) (hornets) . . . .*Vespa* spp.

     (European hornet) *V. crabro*

- Vertex short, distance from lateral ocellus to occipital carina subequal to distance between lateral ocelli (Figure 21.38) 16

16. Oculo-malar space narrow, compound eyes nearly touching base of mandibles (Figure 21.37) (below-ground yellowjackets) . . . .*Vespula* spp.
- Oculo-malar space broad, compound eyes separated from base of mandibles by distance subequal to distance between base of antennae (bald faced "hornet" and above-ground yellowjackets) . . . .*Dolichovespula* spp.

---

[a]Modified from Pratt and Stojanovich (1966); prepared with aid of J. F. MacDonald.
[b]Includes social species that are serious stinging threats owing to large concentrations of individuals.
[c]Adapted from Akre et al. (1981).

References

Akre, R. D., A. Greene, J. F. MacDonald, P. J. Landolt, and H. G. Davis. 1981. *The Yellowjackets of America North of Mexico.* USDA Agr. Handbook 552, 105 pp.

Atkins, J. A., W. C. Wingo, and W. A. Sodeman. 1958. Necrotic arachnidism. *Am. J. Trop. Med. Hyg.* 7:165–84.

Baerg, W. J., 1959. *The Black Widow and Five Other Venomous Spiders in the United States.* University of Arkansas Agr. Exp. Stat. Bull. 608. 43 pp.

Bohart, R. M., and R. C. Bechtel. 1957. *The Social Wasps of California (Vespinae Polistinae, Polybiinae)* Bull. California Insect Survey, vol. 4, no. 3. University of California Press, Berkeley, pp. 73–102.

Byalynitskii-Birulya, A. A. 1917. *Arthrogastric Arachnids of Caucasia. I. Scorpiony.* Anns. Caucasion Mus. A(5). 170 pp. Translated from Russian 1964, Israel Program for Scientific Translations.

Gorham, J. R. 1968. The geographic distribution of the brown recluse spider, *Loxosceles reclusa,* and related species in the U.S. *Coop. Econ. Insect Report. 18:*171–3.

Horen, W. P. 1963. Arachnidism in the United States. *J. Am. Med. Assoc. 185:*839–43.

Kaston, B. J. 1948. *Spiders of Connecticut.* State Conn. Public Doc. 47, Geo. Natural Hist. Survey Bull. 70. 874 pp.

Kaston, B. J., and E. Kaston. 1952. *How to Know the Spiders.* Picture Key Nature Series. Wm. C. Brown, Dubuque, Iowa. 220 pp.

Levi, H. W., L. R. Levi, and H. S. Zim. 1968. *Spiders and Their Kin.* A Golden Nature Guide. Golden Press, New York. 160 pp.

Pratt, H. D., and C. J. Stojanovich. 1966. Hymenoptera: Key to some common species which sting man. In *Pictorial Keys to Arthropods, Reptiles, Birds and Mammals of Public Health Significance.* U.S. Dept. of Health, Education and Welfare, Communicable Disease Center, Atlanta, pp. 102–18.

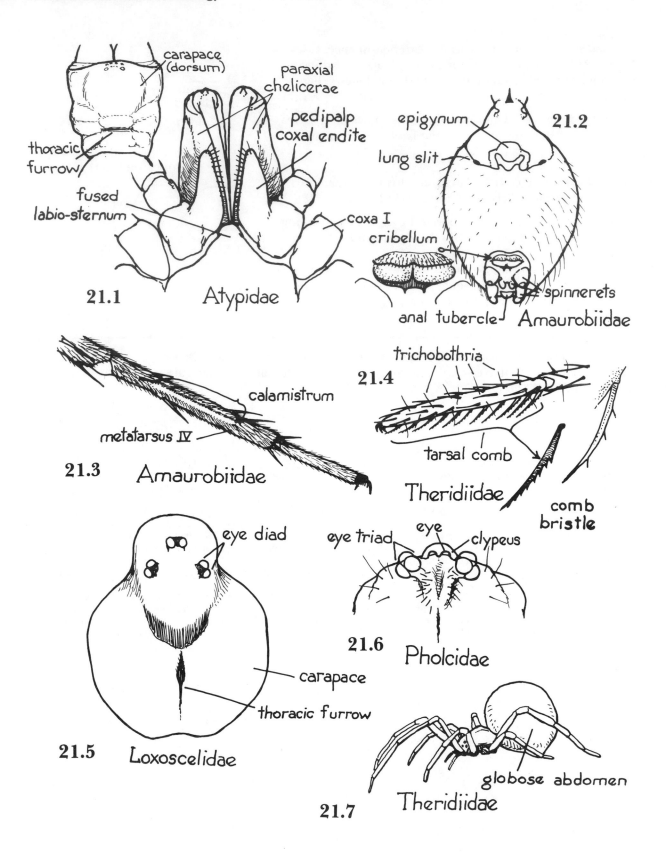

21.1   Atypidae

carapace (dorsum)

paraxial chelicerae

pedipalp coxal endite

thoracic furrow

fused labio-sternum

coxa I

21.2   Amaurobiidae

epigynum

lung slit

cribellum

spinnerets

anal tubercle

21.3   Amaurobiidae

calamistrum

metatarsus IV

21.4   Theridiidae

trichobothria

tarsal comb

comb bristle

21.5   Loxoscelidae

eye diad

carapace

thoracic furrow

21.6   Pholcidae

eye triad

eye

clypeus

21.7   Theridiidae

globose abdomen

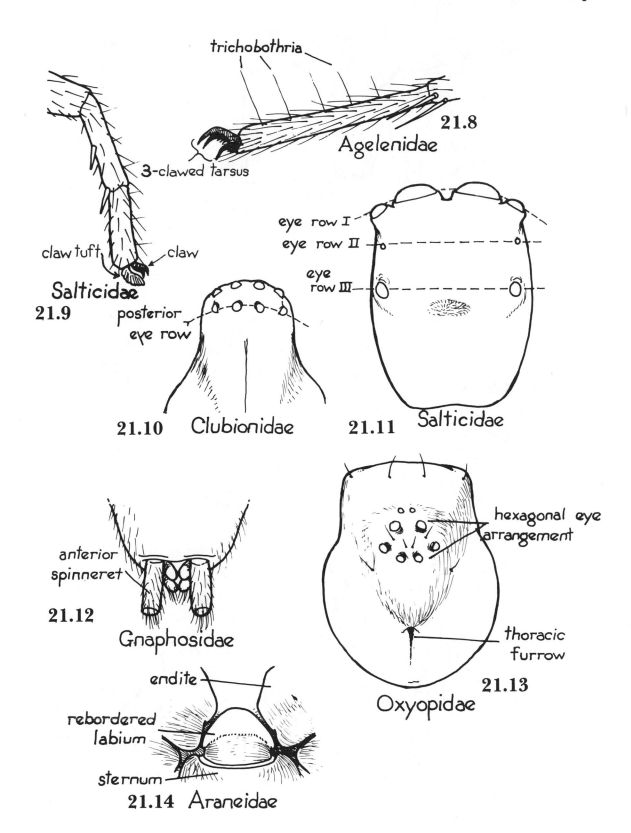

trichobothria

21.8
Agelenidae

3-clawed tarsus

claw tuft   claw

Salticidae

21.9

posterior
eye row

eye row I
eye row II

eye
row III

Salticidae

21.10 Clubionidae   **21.11**

anterior
spinneret

hexagonal eye
arrangement

21.12
Gnaphosidae

thoracic
furrow

21.13
Oxyopidae

endite

rebordered
labium

sternum

21.14 Araneidae

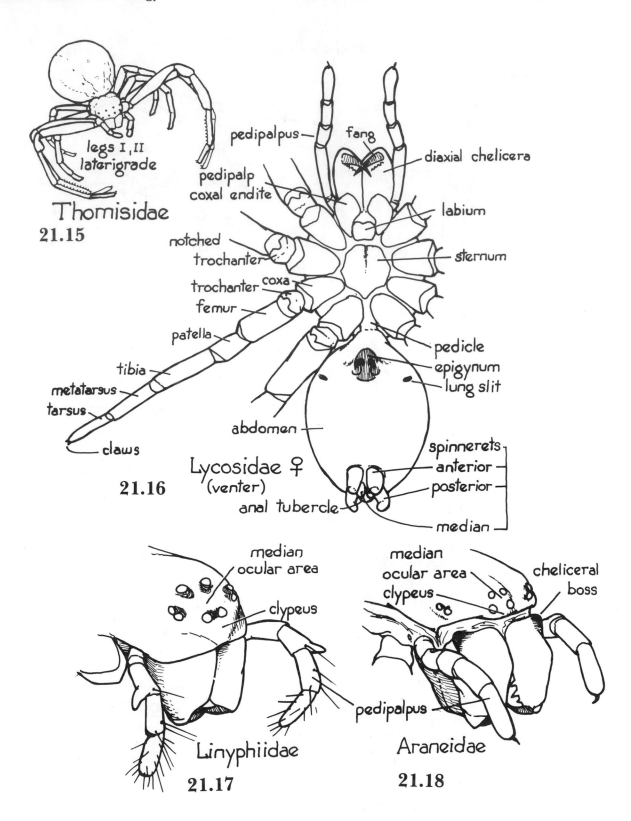

legs I, II
laterigrade

Thomisidae
21.15

pedipalpus — — fang

pedipalp
coxal endite

— diaxial chelicera

notched
trochanter

— labium

trochanter — coxa

— sternum

femur

patella

— pedicle
— epigynum
— lung slit

tibia

metatarsus

tarsus

claws

abdomen

Lycosidae ♀
(venter)
21.16

spinnerets
anterior —
posterior —
median —

anal tubercle —

median
ocular area

— clypeus

Linyphiidae
21.17

median
ocular area
clypeus

cheliceral
boss

pedipalpus

Araneidae
21.18

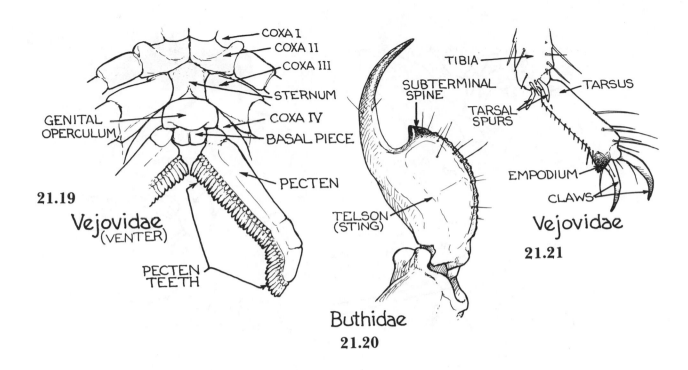

COXA I
COXA II
COXA III
STERNUM
GENITAL OPERCULUM
COXA IV
BASAL PIECE
PECTEN

21.19
Vejovidae
(VENTER)

PECTEN TEETH

TIBIA
SUBTERMINAL SPINE
TARSAL SPURS
TARSUS
EMPODIUM
CLAWS

TELSON (STING)

Vejovidae
21.21

Buthidae
21.20

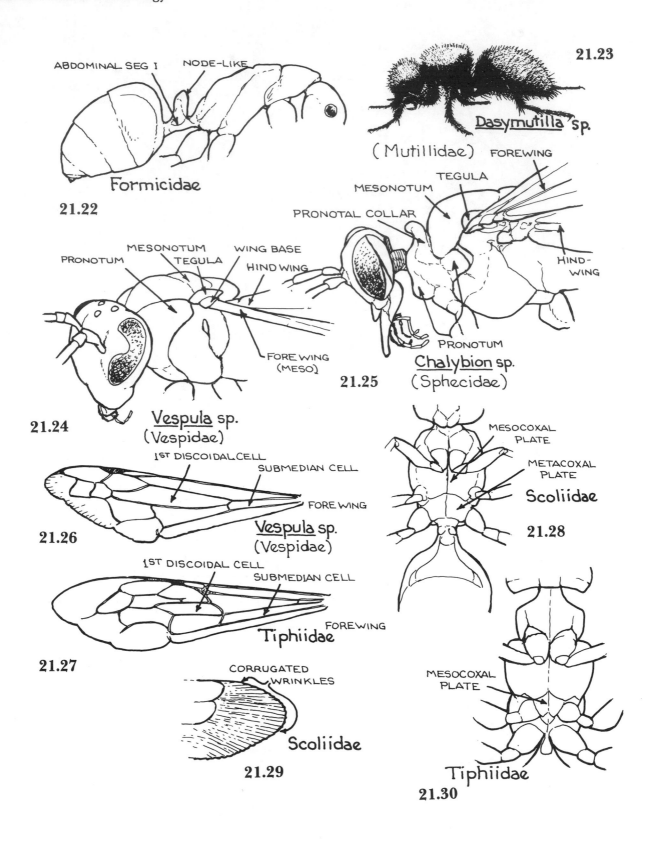

ABDOMINAL SEG I   NODE-LIKE

Formicidae

21.22

21.23

Dasymutilla sp.

(Mutillidae)

MESONOTUM   TEGULA   FOREWING

PRONOTAL COLLAR

HIND-WING

PRONOTUM

Chalybion sp.

(Sphecidae)

21.25

PRONOTUM   MESONOTUM   WING BASE

TEGULA   HIND WING

FORE WING
(MESO)

Vespula sp.
(Vespidae)

21.24

1ST DISCOIDAL CELL

SUBMEDIAN CELL

FORE WING

Vespula sp.
(Vespidae)

21.26

MESOCOXAL
PLATE

METACOXAL
PLATE

Scoliidae

21.28

1ST DISCOIDAL CELL

SUBMEDIAN CELL

Tiphiidae   FOREWING

21.27

CORRUGATED
WRINKLES

Scoliidae

21.29

MESOCOXAL
PLATE

Tiphiidae

21.30

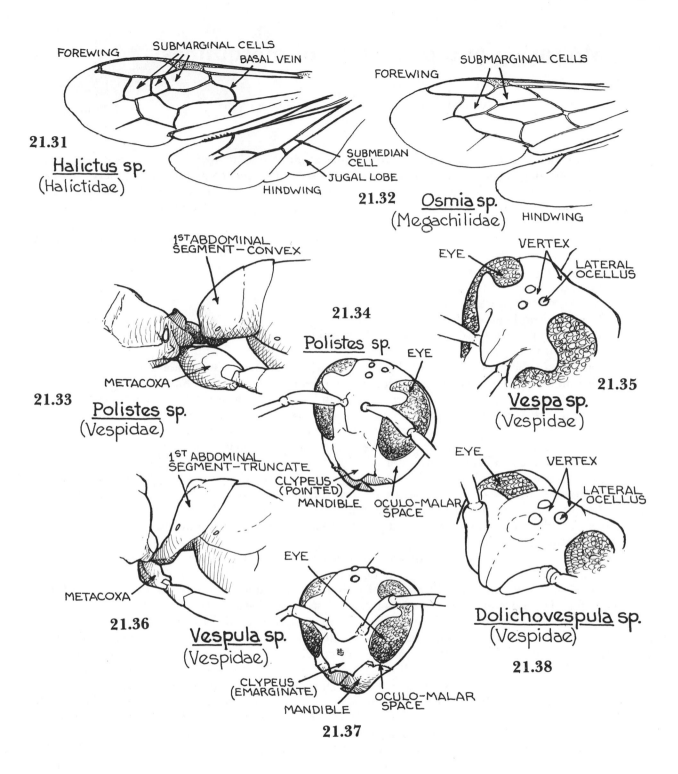

21.31 <u>Halictus</u> sp. (Halictidae)

FOREWING
SUBMARGINAL CELLS
BASAL VEIN
SUBMEDIAN CELL
JUGAL LOBE
HINDWING

21.32 <u>Osmia</u> sp. (Megachilidae)

FOREWING
SUBMARGINAL CELLS
HINDWING

21.33 <u>Polistes</u> sp. (Vespidae)

1ST ABDOMINAL SEGMENT – CONVEX
METACOXA

21.34 <u>Polistes</u> sp.

EYE
CLYPEUS (POINTED)
MANDIBLE
OCULO-MALAR SPACE

21.35 <u>Vespa</u> sp. (Vespidae)

EYE
VERTEX
LATERAL OCELLUS

21.36

1ST ABDOMINAL SEGMENT – TRUNCATE
METACOXA

21.37 <u>Vespula</u> sp. (Vespidae)

EYE
CLYPEUS (EMARGINATE)
MANDIBLE
OCULO-MALAR SPACE

21.38 <u>Dolichovespula</u> sp. (Vespidae)

EYE
VERTEX
LATERAL OCELLUS

22   Mounting and dissection of mosquitoes

In Exercise 3 you learned general mounting techniques applicable to arthropods of medical importance. Because mosquitoes play a role of major importance as pests and vectors of pathogenic agents of man and other animals, it is essential that certain special techniques concerned with their mounting be considered. Dissection methods for adult mosquitoes also are frequently necessary to determine the presence or absence of oocysts or sporozoites.

Larval mosquitoes may be killed with hot water and preserved in 70% ethyl alcohol. In field collecting, larvae may be placed directly in 95% alcohol because some water is normally added with the larvae. A rapid mounting technique is as follows:

1. Ten to 15 minutes in 70% ethyl alcohol
2. Five minutes in 95% alcohol
3. Five seconds in absolute alcohol
4. Transfer to slide and clear in U.S.P. creosote for a few minutes
5. Mount in balsam

The posterior tip of the abdomen of culicine larvae should be cut part way through dorsally between the 7th and 8th segments so that the anal segment and siphon may be oriented with the lateral side up on the finished mount. This should be done while the larva is in 70% alcohol as it will become increasingly brittle when dehydrated. For other mounting procedures, see the references given at the end of the exercise.

In recent years, interest has increased in the identification of pupal stages of mosquitoes. Pupae are best studied by mounting the cast exuvia, or pelts, after adult mosquitoes have emerged.

Exuvia will deteriorate rapidly if left in water, and for this reason should be placed in 70% or 95% alcohol as soon as possible after the adult emerges. The exuvium or pelt is teased apart between the meso- and meta-thoracic sections so that it is separated into an anterior thoracic unit and a posterior abdominal and meta thoracic unit. Both units are mounted on a slide following the technique described for larvae.

Laboratory-reared associations of larval and pupal exuvia belonging to an individual adult mosquito may be mounted together on the same slide.

Often, detailed study of the male genitalia is necessary for the identification of closely related adult mosquitoes. Such study usually requires special preparation. Cut off the posterior one-fourth of the abdomen of a relaxed or freshly killed mosquito and place it in a 10% solution of potassium hydroxide in a small porcelain crucible. Heat almost to the boiling point. Rinse specimen in several changes of water over a 10- to 15-minute period, then transfer it to a drop of water on a microscope slide. Under the dissecting microscope free the genitalia from the preceding segments. Remove the debris, leaving only the genitalia. Draw off the water with blotting paper, replacing with 95% alcohol, changing alcohol twice at 1-minute intervals; replace with absolute alcohol for 1 minute, then with xylol. Orient the specimen, spreading with dorsal side up, drain it, and mount in balsam using a small cover glass. Alternatively, after drawing the water off the specimen, a drop of chloral gum mounting medium may be applied, applying the cover slip directly. The cover slip should be supported on small pieces of broken cover glass to avoid crushing the specimen.

Dissection of adult mosquitoes for malaria parasites is best made on a microscope slide in a drop of physiological saline under a medium-power dissecting microscope. Only freshly killed mosquitoes should be used. A single specimen may be used for examination of the stomach for oocysts, and of the salivary glands for sporozoites of malaria plasmodia. The equipment needed is as follows: two fine dissecting needles, one pair fine forceps, physiological saline (0.85% NaCl) solution tinted with methylene blue.

Wings and legs of the mosquito are cut off close to the body, and the specimen is placed on a drop of tinted saline on a microscope slide. The insect is oriented on its right side with the abdomen toward the observer. The integument on dorsal and ventral surfaces is nicked between the 5th and 6th abdominal segments (Figure 22.1). Hold the thorax with one needle and gently pull on the caudal abdominal segments with the other needle; the gut will be pulled free of the body, usually breaking anterior to the midgut. Arising from the posterior end of the midgut will be seen the Malpighian tubules. These should be cut short, as should the hindgut just behind the stomach. The stomach may then be transferred to a drop of saline on a fresh slide for examination under the compound microscope.

To isolate the salivary glands the mosquito is oriented with the head toward the observer. Gentle pressure is exerted on the thorax with the side of a needle so that the neck becomes distended. With another needle pull the head from the body, while maintaining pressure on the thorax (Figure 22.1). The salivary glands will usually be forced out of the thorax and remain attached to the head. The paired glands are visible as small, trilobed, multicellular organs, more intensely blue-stained than surrounding tissues. If they break off and remain in the thorax, they may be dissected free from their position just above the insertion of the first pair of legs. The glands may be placed in a fresh drop of tinted saline on a second slide for examination with the compound microscope.

Methods for preparing permanent, stained mounts of salivary glands and stomach with associated parasites are described by Boyd (1949) and Russell et al. (1963).

The age grouping of Diptera

Generally hematophagous insect vectors must blood-feed at least twice in order to acquire and subsequently transmit a pathogen. In some Diptera, such blood meals are necessary for the completion of egg development, owing to the protein requirements of ovogenesis. Except for species or individuals displaying autogeny, females rarely lay more egg batches than the number of complete blood meals taken. Thus in the sampling of a Diptera population, if the mean number of egg batches laid per female (i.e., number of gonotrophic cycles completed) can be determined, then the mean number of blood meals per female is assumed to be similar. Knowing such information is important, if we wish to learn the vector potential of a given species in an epidemic. A complication to this assumption is that some individuals will take on several partial blood meals in completing one gonotrophic cycle.

Changes in the tracheal system of the ovaries and midgut are irreversible, making it possible to distinguish parous from nulliparous females. Examination of the trachea of the ovary and of the midgut is the method used. Once these organs have been extracted from the same female, their condition can aid in establishing whether a particular race or species displays autogeny (i.e., if examination of the ovarial tracheal system indicates the female has oviposited, and examination of the tracheal system of the midgut indicates she has not blood-fed, autogeny can be assumed). If a number of females show such characteristics, then that race or species to which the females belong probably is autogenous.

Determination of the physiological age of a female mosquito can be made accurately by examining individual ovarioles for the number of dilatations present. For each gonotrophic cycle, each ovariole produces a new follicle which may develop to maturity or have its development arrested (Detinova, 1962).

Autogeny complicates such a survey in that an autogenous female may have oviposited twice, but only blood-fed once. In such a case, she is not an infective vector unless pathogens, such as some arboviruses, were acquired through transtadial passage.

Indirect determination of the number of gonotrophic cycles is the objective of physiological age grouping. The following laboratory exercise will deal with qualitative and quantitative means of determining the physiological age development of one medically important family of Diptera, Culicidae. Detinova (1968) reviews age structure analysis and physiological "aging" for several groups of Diptera, and should be consulted as a starting point for further information. These techniques need not be restricted to blood-sucking insects, but can be used for aging potential vectors in general (Anderson, 1964).

Materials and methods

Techniques for examining three physiological changes will be discussed in this exercise: (1) tracheation of the midgut, (2) tracheation of the ovaries. and (3) determination of the number of dilatations in individual ovarioles. Other age-estimating methods, such as external physical wear (Corbet, 1960), phragma growth (Schlein and Gratz, 1973), or abdominal wall pigmentation (Dyce, 1969), are not included here.

Tracheation techniques

Materials needed for the two tracheation techniques are simple and easily obtainable: (1) minuten probes, (2) ether, (3) alcohol (70–95%), (4) a dissecting scope, and (5) microscope slides.

For these procedures the best results will be obtained with freshly killed adult female mosquitoes. In most cases this requires maintenance of a laboratory colony (Exercise 24). *Culex pipiens* is a species that can be maintained readily through several ovogenic cycles in a laboratory colony. In this way individual females with a known history of blood feeding and oviposition may be examined. Stored specimens, frozen submerged in water, also can be used. The methods for examining the tracheal changes are as follows:

1. Anesthetize the live mosquito with ether.
2. Place the mosquito in alcohol for approximately 30 seconds.
3. Remove the mosquito from alcohol and blot the excess.
4. Place the mosquito in a few drops of physiological saline.
5. With the dissecting scope on low power, dissect the mosquito (against a black background).

The dissection is done as follows: Place one minuten through the thorax, and with the other, pierce the penultimate segment of the abdomen. Now, with a slow and steady pressure, pull the minutens in opposite directions (Figure 22.1). The ovaries and midgut should pull away with the penultimate segment (Figure 22.2). Separate the detached ovaries and midgut from the Malpighian tubules, fat bodies, and so on, that also have been pulled from the body cavity.

Examination of the ovaries and midgut may be made immediately, or they may be allowed to dry on the slide and examined later. No cover slip or other preparation is needed for the dried slides.

Terminal tracheal **skiens,** or knots, indicate that the underlying organ has not enlarged greatly during the life of that individual. Such enlargement would be caused by a midgut distended by a blood meal (nectar meals distend the crop and not the midgut) or by an ovary distended by maturing eggs. Once the organ has enlarged and the skiens have stretched, the trachea will not return to the tightly knotted appearance even though the organ shrinks to its former size. The lack of skiens on midgut or ovaries indicates a past blood meal or previous ovogenesis (i.e., parity). Thus **parous** and **nulliparous** females can be differentiated by examining tracheation of the ovaries (Figure 22.4), and **autogeny** is indicated when skiens are present on the midgut but not on the ovaries.

Ovariolar dilatations

The materials needed for this technique are the same as those used in the tracheal examination techniques, with the addition of crystal violet stain and a compound microscope. The procedure is identical to the one outlined for tracheation. Once the ovaries are free, add a drop of dilute crystal violet stain. Observe the ovaries until they are lightly stained (1–2 minutes) and then blot up the fluid, taking care not to disturb the ovaries.

The following are done under a compound microscope:

1. Immobilize the ovary with one minuten probe, and with the other probe pierce the ovary sheath and remove a section of it to expose the ovarioles.
2. Tease and spread some ovarioles apart without tearing them away from the internal oviduct (Figure 22.3).
3. Choose an ovariole and carefully pull it away from the oviduct to straighten it out and view its entirety. Dilatations (Figure 22.5) can be counted. Dilatations consist of stained "relics" of debris from previous ovogenic cycles that remain lodged in the ovariolar pedicel. Care must be taken not to tear the ovariole from the oviduct in order to be certain that any dilatations present are not lost.
4. To ensure accuracy, a number of ovarioles should be examined and the dilatations counted. Carpenter and Nielsen (1965) examined 10–15 ovarioles per female. There are three reasons for doing so: (1) not every follicle develops for each gonotrophic cycle, (2) there are occasions in which fully developed follicles are not released during oviposition (these eggs are either later released or resorbed), and (3) a few ovarioles may degenerate or cease to function after one or two gonotrophic cycles.

Stage of follicular development

A final factor to consider is the stage of development of the follicle. Carpenter and Nielson (1965) classify the stages as follows:

Stage N. Follicle consists of eight undifferentiated cells.

Stage I. Follicle is spherical or slightly oval without yolk granules. One oocyte and seven nurse cells are present.

Stage II. Follicle is oval with yolk granules occupying less than half of the follicle.

Stage III. Follicle becomes slightly elongated; oocyte fills more than half of the follicle; yolk granules occupy more than half of the follicle.

Stage IV. Follicle becomes distinctly elongate (length-to-width ratio is greater than 2:1); yolk granules occupy almost all of the follicle; nurse cells are only faintly visible.

Stage V. Follicle contains a fully formed egg (length-to-width ratio is greater than 5:1) with visible chorionic structures, ready for oviposition

Relic in pedicle is remnants of nurse cells and follicular epithelium.

Mention is made of these stages because the eggs of a non-blood-fed, nonautogenous female usually will develop to Stage II and then cease to develop until a blood meal is obtained (i.e., *Culex pipiens* will not develop eggs past Stage II without a blood meal). Therefore, a nulliparous female, which does not display autogeny, in an advanced stage of follicular development probably has blood-fed once. Determination of the number of relics and the follicular stage is used to estimate the number of blood meals taken by a female. Whether or not she has blood-fed can be confirmed by examining midgut tracheation.

References

Anderson, J. R. 1964. Methods for distinguishing nulliparous from parous flies and for estimating the ages of *Fannia canicularis* and some other cyclorraphous Diptera. *Ann. Entomol. Soc. Am. 57*:226–36.

Boyd, M. F. 1949. *Malariology*. W. B. Saunders, Philadelphia, pp. 191–5, 350–6.

Carpenter, S. J., and W. J. LaCasse. 1955. *Mosquitoes of North America*. University of California Press, Berkeley.

Carpenter, M. J., and L. T. Nielsen. 1965. Ovarian cycles and longevity in some univoltine *Aedes* species in the Rocky Mountains of the western United States. *Mosquito News 25*:2.

Corbet, P. S. 1960. Recognition of nulliparous mosquitoes without dissection. *Nature 187*:525–6.

Detinova, T. S. 1962. *Age-grouping Methods in Diptera of Medical Importance*. World Health Organization, Geneva. 216 pp.

– 1968. Age structure of insect populations of medical importance. *Ann. Rev. Entomol. 13*:427–50.

Dyce, A. L. 1969. The recognition of nulliparous and parous *Culicoides* (Diptera: Ceratopogonidae) without dissection. *J. Australian Entomol. Soc. 8*:11–15.

Matheson, R. 1944. *The Mosquitoes of North America*. Comstock, Ithaca, N.Y.

Russell, P. F., L. S. West, R. D. Manwell, and G. MacDonald. 1963. *Practical Malariology*. Oxford University Press, New York.

Schlein, Y., and N. G. Gratz. 1973. Determination of the age of some anopheline mosquitoes by daily growth layers of skeletal apodemes. *Bull. W.H.O. 49*:371–5.

22.1
LINE OF EXTERNAL SEPARATION PHARYNGEAL PUMP
SPIRACLE
HOLDING POINT
DIRECTION OF PULL
CIBARIAL PUMP
SALIVARY GLAND
MID GUT
HIND INTESTINE
OVARY
PROBABLE LINE OF INTERNAL SEPARATION
LINE OF EXTERNAL SEPARATION DIRECTION OF PULL
MID GUT
OVARY
MALPIGHIAN TUBULE

22.2
EXPOSED SALIVARY GLANDS
TRACHEA
MID GUT
HIND INTESTINE
OVARY
TERMINAL SEG.

TRACHEAL SKIEN
1 HR POST BLOODMEAL (BM)
24 HR POST BM
48 HR POST BM
22.4 120 HR POST BM

22.3
TRACHEA
SUSPENSORY LIGAMENT
OVARY
OVARIOLE
OVIDUCT
ACCESSORY GLAND
SPERMATHECA
GENITAL ATRIUM

22.5
GERMARIUM
NURSE CELLS
OÖCYTE (3RD CYCLE)
PEDICEL
RELIC MATTER (2ND CYCLE)
PEDICEL
DILATATION
RELIC (1ST CYCLE)

23 Blood pathogen and blood meal identification

Hematology, the study of blood, is a necessary element in the epidemiology of arthropod-borne diseases. Blood can be sampled for study either from the vertebrate host or from the blood-fed arthropod. In addition to normal cellular elements, a blood sample may contain hemopathogens. The medical entomologist should be able to recognize the various normal blood cells as well as arthropod-borne pathogens such as certain Protozoa and Metazoa found in the blood of infected vertebrate hosts. This exercise demonstrates the preparation and staining of a blood smear for microscopic examination, identification of some larger blood pathogens, and the determination of the source of an arthropod blood meal. Before beginning, you should read some concise account of normal blood (e.g., Chapter 14 of Stitt et al., 1948). Techniques and guidelines for procedures used in conducting epidemiological surveys can be found in Sudia et al. (1972).

Preparation and staining of blood smear

Under direction of the instructor prepare several thin blood smears. Smears will be stained with Wright's or Giemsa's stain. Wright's staining solution is prepared by dissolving 0.3 g Wright's dye in 100 ml absolute methyl alcohol. It is important that the alcohol be chemically pure and acetone-free. Filter staining solution before use. The dye may be prepared in the laboratory (see Garcia and Ash, 1979) or purchased ready for use as a liquid.

Using Wright's stain, the air-dried blood film should be covered with about 10 drops of stain and allowed to stand for 1 minute. An equal quantity of distilled water or buffered diluent then is added dropwise. (Buffer is composed of 0.63 g monobasic potassium phosphate, 2.56 g anhydrous dibasic sodium phosphate, and 1000 ml distilled water.) Allow blood film to stand 4 or 5 minutes; then wash it with running distilled water and air dry.

Stock Giemsa stain is prepared by dissolving 0.8 g Giemsa powder (National Aniline No. 561) in 50 ml of glycerin, 95%, C. P., and 50 ml absolute methyl alcohol. Two to 3 days of shaking on a mechanical shaker are required to dissolve the stain, after which it is filtered. Giemsa stain is available commercially as a stock solution. Just prior to use, the stock solution is diluted with buffered distilled water (pH 6.51) at the rate of 1 part of stain to 49 parts water.

Before using Giemsa's stain, fix the blood film in absolute methyl alcohol for 1 minute and air dry. Cover it with diluted stain for 15 to 40 minutes. Wash it briefly in distilled water until the film has a pinkish tinge. Air dry.

Identification of pathogens in blood smear

Examine stained smears of normal blood, noting the characteristic appearance of erythrocytes, blood platelets, neutrophils, eosinophils, basophils, lymphocytes, and monocytes.

Study stained thin blood films showing the three common species of human malaria plasmodia. Compare with the excellent color plates of malaria plasmodia given by Wilcox (1960), Garnham (1966), Faust et al. (1970), or Garcia and Ash (1979). Sketch representative stages in development of the parasites. Drawings should be arranged in three parallel columns, so that the various corresponding stages will appear in a row across the page.

Plasmodium vivax
1. Infected cell larger than normal, often misshapen, frequently with red stippling (Schüffner's dots) (Figure 23.1). May be two or three parasites in a cell.
2. Ring stage (early trophozoite). Heavy chromatin dot and large cytoplasmic circle (Figure 23.1).

3. Large trophozoite.
4. Mature parasite (schizont). Twelve to 24 divisions, usually 16 (merozoites), each containing a chromatin dot and cytoplasm (Figure 23.1).
5. Sexual stages. Circular or ovoid outline. Microgametocyte does not fill enlarged erythrocyte; cytoplasm pale; chromatin light red or pink, large and diffuse. Macrogametocyte fills enlarged erythrocyte; cytoplasm dark blue, lacking vacuoles; chromatin deep red, small and compact; pigment granules abundant, dark brown.

Plasmodium malariae
1. Infected cell of normal size, rarely containing more than one parasite.
2. Ring stage. Cytoplasmic circle usually small, thick; chromatin dot heavy.
3. Large trophozoite. Cytoplasm compact. Oblong or band shaped with few irregularities.
4. Mature schizont. Six to 12 merozoites, usually 8, practically filling red blood cell.
5. Sexual stages. Microgametocyte similar to *vivax* but infected cell not enlarged. Macrogametocyte similar to *vivax* but pigment granules coarser; infected cell not enlarged.

Plasmodium falciparum
1. Infected cell of normal size, frequently infected with more than one parasite.
2. Ring stage. Cytoplasmic ring delicate, small; one or two small chromatin dots.
3. Large trophozoite. Rarely seen in peripheral blood. Small, about half the diameter of infected cell. Cytoplasm compact, lightly staining. One small chromatin mass.
4. Mature schizont. Rarely seen in peripheral blood. Fills about two-thirds of infected cell; 8 to 24 small merozoites.
5. Sexual stages. Sausage- or crescent-shaped. Microgametocyte grayish-blue to pink with diffusely scattered red chromatin. Macrogametocyte a deeper blue with a dense, red chromatin mass; dark pigment granules more concentrated than in microgametocyte.

Examine a stained blood smear (Romanowsky type of stain) from cattle infected with *Anaplasma marginale*. Note the small size (0.5 μm diameter) of the spherical or oval marginal bodies in erythrocytes. Draw several infected cells. (The protozoan nature of the marginal anaplasma bodies is questionable. They have been reported to have structural characteristics similar to those of inclusion bodies of certain viruses, as well as to the initial bodies of rickettsiae.)

Examine a Giemsa-stained film of bone marrow containing *Leishmania donovani*. Observe the ovoidal or rounded bodies (2-3 μm long) appearing in infected reticulo-endothelial cells. Note the pale blue cytoplasm, the red-stained nucleus, and the deep red or violet, rodlike kinetoplast. In well-stained specimens the dotlike basal body may be visible near the kinetoplast. Draw representative specimens.

Examine a Wright- or Giemsa-stained blood film showing *Trypanosoma cruzi* (Figure 23.3). Draw a representative form showing the spindle shape, the rounded, centrally placed nucleus, the posterior oval kinetoplast, and the dotlike basal body, from which arises the axoneme which borders the undulating membrane and projects anteriorly as a free flagellum. Observe a demonstration slide showing living trypanosomes in a fecal smear from a *Triatoma*.

Examine stained and/or fresh smears containing *Borrelia* (e.g., *B. recurrentis* from infected mouse blood or *Borrelia.* spp. from intestines of fleas, cockroaches, sheep keds, or termites). Note size, shape, and, in living specimens, the characteristic movements of the *Borrelia* (Fig. 23.2).

Examine stained blood films containing microfilariae of *Wuchereria bancrofti*, *Dirofilaria immitis*, *Litomosoides carinii*, or other related filarial nematodes. (Figure 23.4). Draw a representative specimen, noting the body length, presence or absence of a sheath, and arrangement of nuclei in tail region. Examine a wet blood smear containing living microfilariae of *D. immitis* or *L. carinii*, noting the characteristic motion of the organisms.

Key to genera of some medically important microfilariae found in peripheral blood or skin

1. Larva enclosed in a transparent sheath (Giemsa stain preparation) 2
• Larva not enclosed in a transparent sheath 4

2. Nuclei extend to caudal tip; body curves kinked 3
- Nuclei lacking in caudal tip; body curves sinuous; length 225–
  300 × 8–10 μm; nocturnal or nonperiodic; in blood or lymph;
  nuclei-free portion of cephalic end subquadrate . . . .*Wuchereria*

3. Length and width of nuclei-free portion of cephalic end
   subequal; length 250–300 × 6–9 μm; diurnal periodic; in
   blood . . . .*Loa*
- Length of nuclei-free portion of cephalic end greater than
  width; length 160–260 × 5–6 μm; nocturnal or subperiodic; in
  blood or lymph; 2 nuclei in caudal tip . . . .*Brugia*

4. Caudal tip sharply pointed; caudal nuclei lacking 5
- Nucleated caudal tip ending bluntly; nonperiodic; length
  180–200 × 4 μm; in skin and blood . . . .*Dipetalonema*

5. From skin or blood of humans or cattle 6
- From blood of Carnivora (dog, fox, raccoon, mink, etc.) . . . .*Dirofilaria*

6. From skin or lymphatics of humans or cattle; nonperiodic;
   length 300–500 × 5–9 μm . . . .*Onchocerca*
- From human blood; nonperiodic; length 170–240 × 5 μm . . . .*Mansonella*

Arthropod blood meal identification

Determination of the feeding habits, or source of blood meal, for hematophagous arthropods is an important element in the epidemiology of arthropod-borne diseases. Identification of vertebrate blood contained in the midgut of a potential vector indicates "host preferences" and possible reservoir sources of blood pathogens. Direct identification of the blood source is possible when the arthropod is observed in the act of feeding. However, techniques used for indirect identification of a given bloodmeal also should be sensitive enough to differentiate one host blood source from another. Cytological characteristics of vertebrate blood cells in the midgut may indicate in a crude way the blood meal source. Nonnucleated red blood cells (RBCs), for example, can only be from a mammalian blood source. Thus diameter of RBCs and the size and shape of the cell nucleus, when present, aid further in crude determination of arthropod blood source.

Two indirect, general methods are used to identify specifically the source of a blood meal; both are based on the unique character of cellular and plasma proteins in the blood of different animals. One method is serological identification, which relies on an antigen (Ag)–antibody (Ab) interaction to produce an observable reaction (e.g., precipitation). The precipitin test (the most used method of choice), the fluorescent Ab test (FA), and the hemagglutination inhibition test (HI) are examples of the serological method. An example of the precipitin test is the agar–gel diffusion technique in which Ag and sera containing Ab are allowed to diffuse through an intervening milieu of gel agar. Where the Ag and Ab meet in proper proportions, a visible line of precipitate forms, indicating a positive reaction.

The second general method is hemoglobin crystallization, which utilizes the formation of crystals of unique configuration following combination of hemoglobin from lysed RBCs with certain chemical reagents. The crystals formed from an unknown blood source are matched with those from a known or reference source.

Both the gel diffusion precipitin test and the hemoglobin crystallization test can be performed in the laboratory to identify blood meals of mosquitoes as described in the following procedures. An active adult mosquito laboratory colony (e.g., *Aedes aegypti*, *Culex pipiens*, or *Anopheles quadrimaculatus*), laboratory mammal blood meal donors, and mammal holding facilities are necessary for these tests.

Mosquito blood meal identification: gel diffusion technique

Principle

Gel diffusion (also known as immunodiffusion or double diffusion) involves Ag/Ab precipitation in a semisolid medium such as agar-gel, starch-gel, or acetate. The object is to allow slow movement (diffusion) of Ag and Ab in opposing directions through this medium so that they reach optimal concentrations and form visible lines of precipitate. This technique gives highly specific results but often requires multiple testing to eliminate cross reactions, and it is difficult to do in the field. Further, because some mosquitoes excrete blood fluid elements soon after feeding, serological techniques, such as this, are not always reliable for determining blood meals several days old.

Procedure

The test requires preparation of antisera (As), preparation of gel plates, and preparation of unknown blood meal (Ag) before doing the test, and performing and reading the gel diffusion test.

Preparation of antisera

Preparation of antisera should be started at least 2 weeks prior to the test.

1. Collect whole blood samples (each 20 ml or more) from a number of animal species on which the laboratory mosquitoes will feed (e.g., chickens, laboratory rat, laboratory mouse, guinea pig, dog).
2. Centrifuge the blood samples (10 minutes at 3000 rpm), and filter the supernatant (Switz Filter pad) using a vacuum flask. Filtering will remove most microbial contaminants which interfere with storage qualities of the blood sample.
3. Inject (1.5 ml, intramuscularly) the supernatant (Ag) into a suitable laboratory animal (e.g., rabbit) and repeat injection with one or two immunizing boosters at 5-day intervals following the initial injection. A different rabbit must be used for each antisera (As) preparation.
4. After 12 to 15 days (following initial injection) recover antiserum as the supernatant of centrifuged whole blood from the rabbit and store it ($-20°C$) until needed for the identification test.

Preparation of gel plates

1. Warm a 1.5% solution of Noble Special agar in buffered physiological saline (0.7 g agar, 0.8 g NaCl dissolved in 100 ml 0.01 molar phosphate-buffered H_2O).
2. Add Merthiolate antiseptic (1:10,000) to the agar mixture after its removal from heat. This will prevent bacterial contaminant growth on the agar. Allow cooling to 60–70°C, and pour warm agar in 8.5-ml lots into each of a series of disposable Petri dishes (100 × 10 mm). Allow the agar to harden with the dish lids off. Cooled agar plates, covered with lids, may be refrigerated for storage until needed.
3. Prior to testing, cut a series of 3-mm-diameter wells in a perimeter around a 3-mm-diameter center well. Use a sterilized cork borer or a gel punch. The resulting pattern will be six or eight peripheral wells surrounding a central well, as shown in Figure 23.6.

Preparation of antigen (unknown blood meal)

1. Anesthetize (CO_2) 20 blood-fed mosquitoes, all of which have fed on the same blood source. Remove and squash their midguts on filter paper. Allow the separate, squashed midgut spots to air dry, and then cut away the excess filter paper. The dried gut–paper discs may be stored ($-20°C$) in a dry sealed container.
2. To ready the gut–paper discs (Ag) for testing, immerse a disc in a 5% solution of physiological saline (0.8 g NaCl in 100 ml H_2O). This will extract the serum proteins (Ag) from the blood spot and hold them in a test solution of saline.

Performing and reading the gel-diffusion test

1. Place a drop of Ag (unknown blood meal suspension) in the center well of a gel plate. Similarly, place drops of known As in the peripheral wells. Use a different pipette for each well loading; use a different plate for each unknown blood meal tested.
2. Cover and incubate the plate at room temperature in a humidified chamber for 24–36 hours.
3. Using a strong narrow beam of light from the side, observe the lines of precipitate in the agar–gel against a black background. Figure 23.5 shows the common types of precipitate reaction that can be observed: negative, positive, nonspecific, and double-lined.

Mosquito blood meal identification: hemoglobin crystallization technique

Principle

Hemoglobin proteins differ among different vertebrate species and thus will produce different crystal forms unique to each species. Hemoglobin crystallization utilizes this formation of characteristic crystals when hemoglobin (from lysed RBCs) reacts with certain **reagents** (e.g., ammonium oxalate). This is a specific, simple, single-step, field-adaptable test. Because it involves blood cells and not serum, the usefulness of this technique increases with older blood meals, but it is not as specific as the precipitin test.

Some other disadvantages of this technique include the ambiguity of some results, the difficulty of obtaining crystallization of bovid hemoglobin, and the difficulty of interpreting results from multiple or mixed blood meals (i.e., where the same midgut contains blood from more than one feeding or blood from different host species).

Procedure

Preparation of test reagent

This test utilizes a buffered (pH 6.8) (phosphate) ammonium oxalate solution (0.035 M). Reagents may be stored (4°C) for no longer than 2 weeks.
1. Stock I: 0.02 M monobasic sodium phosphate (2.78 g Na_2PO_4/100 ml distilled H_2O)
2. Stock II: 0.02 M dibasic sodium phosphate (5.36 g $Na_2PO_4 \cdot 7H_2O$/100 ml distilled H_2O)
3. Reagent buffer: 51 ml Stock I, 40 ml Stock II in 100 ml distilled H_2O = 200 ml 0.1 M Na_2PO_4 buffer at pH 6.8
4. Final reagent: 0.035 M ammonium oxalate (0.434 g ammonium oxalate, 25 ml 0.1 M buffer, distilled H_2O to make 100 ml total solution)

Preparation of unknown blood meal
1. Feed about 20 laboratory colony mosquitoes on each of several restrained hosts (guinea pig, laboratory mouse, and human hosts give good crystal contrast). Feeding is best when done in early morning or evening. Hold the engorged mosquitoes for 6–12 hours (at 22°C).
2. Anesthetize the mosquitoes, and remove the midguts as shown in Exercise 22. Midguts may be stored (at −20°C) in a sealed dry container for later testing.

Hemoglobin crystallization test
1. Place a single midgut in a depression slide or microtiter well and add 10 μl of ammonium oxalate reagent (0.035 M).
2. Macerate the midgut with a small glass rod, and remove the suspension from the depression or microwell with a 32-mm capillary tube.
3. Seal one end of the tube with clay sealer and centrifuge (5 minutes, 3000 rpm) in a micro-hemacrit centrifuge.
4. Cut off the sealed, sediment end of the capillary tube, and place the supernatant on a clean glass slide.
5. Allow the supernatant to dry until a ring forms on the periphery, and then apply a cover slip. Seal the cover slip (Permount) 1–2 hours after covering, and refrigerate (4°C) for storage.
6. Examine peripheral Hb crystals under a compound microscope and match the crystal forms with reference forms (Figure 23.7). Reference crystal slides may be prepared from known whole blood as described in Washino and Else (1972) or Washino (1977).

References

Bull, C. G., and W. V. King. 1923. The identification of the blood-meal of mosquitoes by means of the precipitin test. *Am. J. Hyg. 3*:491–6.

Crans, W. J. 1969. An agar gel diffusion method for the identification of mosquito blood-meals. *Mosquito News 9*:563–6.

Cowle, A. J. 1958. A simplified micro double-diffusion agar precipitin technique. *J. Lab. Clin. Med. 52*:784–7.

Faust, E. C., P. F. Russell, and R. C. Jung. 1970. *Craig and Faust's Clinical Parasitology*, 8th ed. Lea & Febiger, Philadelphia.

Garcia, L. S., and L. R. Ash. 1979. *Diagnostic Parasitology*, 2nd ed. Mosby, St. Louis.

Garnham, P. C. C. 1966. *Malaria Parasites and Other Haemosporidia*. Blackwell, Oxford.

McKinney, R. W., V. T. Spillane, and P. Holden. 1972. Mosquito blood meals: identification by a fluorescent antibody method. *Am. J. Trop. Med. Hyg. 21*:999–1003.

Stitt, E. R., P. W. Clough, and S. E. Branham. 1948. *Practical Bacteriology, Hematology, and Parasitology,* 10th ed. Blakiston, Philadelphia.

Sudia, W. D., R. D. Lord, and R. O. Hayes. 1972. *Collection and Processing of Vertebrate Specimens for Arbovirus Studies.* U.S. Dept. of Health, Education and Welfare, Pub. Health Serv., Center for Disease Control, Atlanta. 65 pp.

Tempelis, C. H. 1975. Host feeding patterns of mosquitoes, with a review of advances in analysis of blood meals by serology. *J. Med. Entomol. 11*:635–53.

Tempelis, C. H., and M. F. Lofy. 1963. A modified precipitin method for identification of mosquito blood-meals. *Am. J. Trop. Med. Hyg. 12*:825–31.

Tempelis, C. H., and M. L. Rodrick. 1972. Passive hemagglutination inhibition technique for the identification of mosquito blood meals. *Am. J. Trop. Med. Hyg. 21*:238–45.

Washino, R. K. 1977. *Identification of Host Blood Meals in Arthropods* (Final Report). U.S. Army Medical Research and Development Command, Washington, D.C. 155 pp.

Washino, R. K., and I. G. Else. 1972. Identification of blood meals of hematophagous arthropods by the hemoglobin crystallization method. *Am. J. Trop. Med. Hyg. 21*:120–2.

Wilcox, A. 1960. *Manual for the Microscopical Diagnosis of Malaria in Man.* U.S. Public Health Service Publ. No. 796. Washington, D.C.

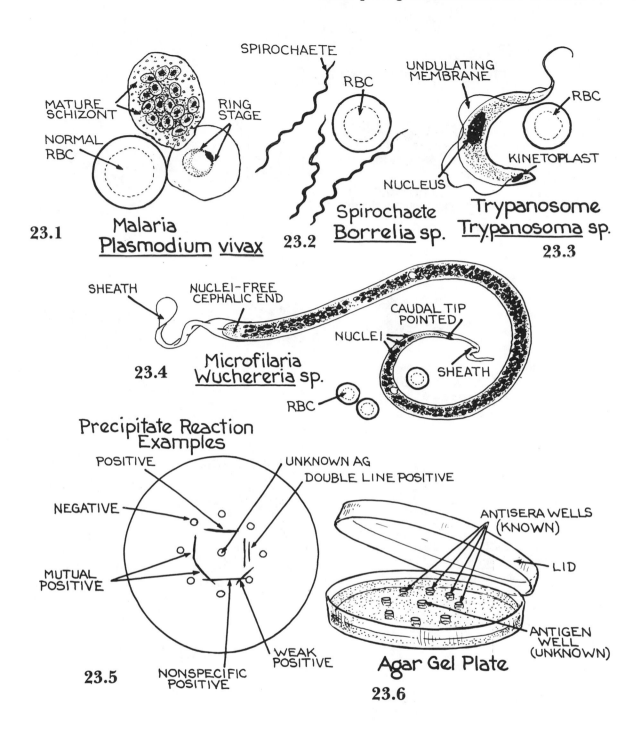

23.1 Malaria
Plasmodium vivax

MATURE SCHIZONT
NORMAL RBC
RING STAGE

23.2 Spirochaete
Borrelia sp.

SPIROCHAETE
RBC

23.3 Trypanosome
Trypanosoma sp.

UNDULATING MEMBRANE
RBC
KINETOPLAST
NUCLEUS

23.4 Microfilaria
Wuchereria sp.

SHEATH
NUCLEI-FREE CEPHALIC END
CAUDAL TIP POINTED
NUCLEI
SHEATH
RBC

Precipitate Reaction Examples

POSITIVE
NEGATIVE
UNKNOWN AG
DOUBLE LINE POSITIVE
MUTUAL POSITIVE
WEAK POSITIVE
NONSPECIFIC POSITIVE

23.5

Agar Gel Plate

ANTISERA WELLS (KNOWN)
LID
ANTIGEN WELL (UNKNOWN)

23.6

23.7
Hemoglobin Crystal Forms:

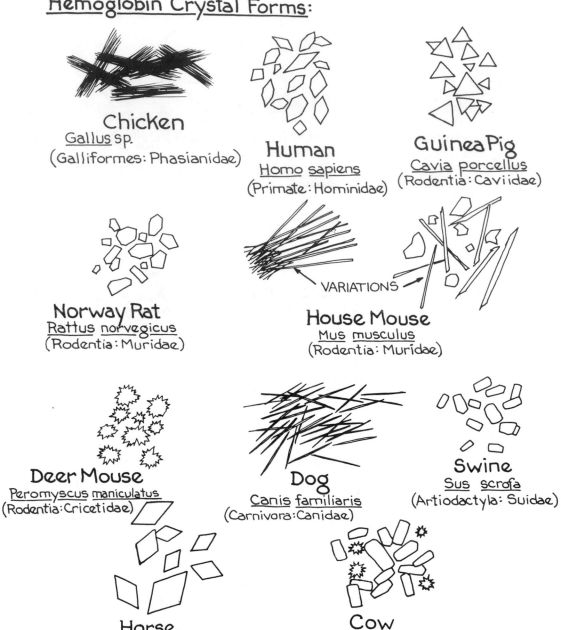

Chicken
Gallus sp.
(Galliformes: Phasianidae)

Human
Homo sapiens
(Primate: Hominidae)

Guinea Pig
Cavia porcellus
(Rodentia: Caviidae)

Norway Rat
Rattus norvegicus
(Rodentia: Muridae)

House Mouse
Mus musculus
(Rodentia: Muridae)

VARIATIONS

Deer Mouse
Peromyscus maniculatus
(Rodentia: Cricetidae)

Dog
Canis familiaris
(Carnivora: Canidae)

Swine
Sus scrofa
(Artiodactyla: Suidae)

Horse
Equus caballus
(Perissodactyla: Equidae)

Cow
Bos taurus
(Artiodactyla: Bovidae)

24 Rearing arthropods of medical importance

Successful rearing of arthropods of medical importance is essential to effective research on their biology, control, and pathogen-vector potentials. Even the identification of specimens may depend upon rearing field-collected material to instars more readily recognized, or previously unassociated in the literature. Cast skins of the various instars may be preserved with the reared adult to validate the association.

Many arthropods are easily reared in the laboratory. Examples include several species of domiciliary cockroaches, house flies, blow flies, some species of mosquitoes, fleas, and ticks. Under optimum conditions the period of time required to rear such arthropods, through a complete cycle from the adult of one generation to that of the next, varies: 10 to 14 days for insects such as house flies and some mosquitoes, a month or more for fleas, 6 months to a year or more for some species of cockroaches, and 2.5 months to 2–3 years for ticks.

One of the first problems to be solved in rearing arthropods collected in the field is that of temporary storage and transportation to the laboratory. Heat and desiccation are rapidly fatal to most arthropods; hence it is essential to have some means of holding living specimens in a relatively cool, moist environment. The details and complexity of the necessary equipment vary with the biology and spatial requirements of the organisms. It may be as simple as a plastic vial containing paper toweling carried in a shirt pocket, as described by Ryckman and Ryckman (1966) for carrying reduviid bugs, with the perspiration of the body providing the needed cooling protection. Many kinds of fleas, mites, or ticks may be carried in stoppered vials and stored temporarily in an ice-filled thermos bottle. The addition of moistened filter paper in vials, although helpful in preventing desiccation of some species, can cause drowning of others, which become trapped in water of condensation on inner surfaces of the vials. A generally safer procedure is to prepare vials in advance of a collecting trip by pouring into each one a bottom layer of a 1:9 mixture of charcoal powder and (gypsum) plaster of Paris mixed to a thick slurry with water. Such vials may be stoppered with simple cork stoppers or with one-hole stoppers in which the hole is covered by a fine mesh cloth. A few drops of water added occasionally to the hardened layer in the bottom of a vial will permit the slow release of water vapor in the bottom of the vial.

Adult Diptera such as synanthropic flies, mosquitoes, and tabanids may be transported in cloth cages supported on wire or wooden frames, the whole covered with toweling or similar material kept saturated with water.

Cockroach rearing

Most domiciliary cockroaches are easily reared arthropods that require relatively little time for maintenance. In general, temperatures of 70–85°F (21–35°C) and relative humidities ranging from 50 to 80% are satisfactory. Food requirments of the common pest species are met by dried, compressed dog chow. Water may be supplied by a variety of devices designed to provide an absorbant moist surface. An effective device consists of a glass tube of water stoppered with a cotton plug and laid on its side in the rearing container. The latter may be a simple 1-gallon glass jar containing a 3- to 5-cm layer of sawdust and a loose roll of corrugated cardboard for harborage. Vaseline smeared inside the inner surface of the top of the jar prevents cockroaches from crawling out of the jar. A cloth cover is fastened over the top of the assembly, or it is closed with a screw-on metal lid in which the center has been replaced with fine mesh screen. Detailed information on rearing cockroaches has been published by Smittle (in Smith, 1966), Tarshis (1961), and others. Handling of cockroaches as well as many other kinds of arthropods is facilitated by chilling, as in a refrigerator, or by anesthetizing them with carbon dioxide gas.

House fly rearing

Laboratory colonization of house flies, *Musca domestica*, is a long-established art in numerous laboratories. Many use a standardized CSMA (Chemical Specialties Manufacturers Association) medium for rearing the larvae. It contains wheat bran, alfalfa meal, yeast, malt, and spent brewer's grains. Various other larval rearing media have been described (see the reviews by Spiller, 1966; Sawicki and Holbrook, 1961; Schoof, 1964). Small-scale rearing of house fly larvae can be accomplished with such materials as fermenting dog biscuits, or laboratory mouse chow, substances that also serve for the successful rearing of some kinds of blow flies. Adult house flies should be provided with a dilute sugar solution continuously available through a reservoir-and-wick arrangement or a similar self-feeding device. Granulated or cube sugar may be fed as a maintenance diet for adult flies, but when oviposition is desired, the diet should be supplemented. An effective oviposition diet consists of a 1:1 dry mix of dried milk and cane sugar supplemented with 10% autolyzed yeast and 0.1% cholesterol. Details of cage construction and handling of reared flies are described in the reviews cited above.

Mosquito rearing

Laboratory culture of mosquitoes is relatively simple for species such as *Culex pipiens*. Others such as *Aedes dorsalis* may require induced mating techniques, long light cycles, or other special conditions which complicate the rearing procedure.

In rearing *C. pipiens* or other species of similar requirements, the eggs, larvae, and pupae are held in shallow white enamel pans containing water to a depth of about 5 cm. Larvae will feed satisfactorily on finely ground dry dog chow. At a water temperature of 27°C hatching of eggs requires about 30 hours, the larval period 6 to 9 days, and the pupal period about 36 hours. Pupae are removed daily, using wide-mouth pipettes, and transferred to water containers placed inside cages to await emergence. Adults should be held in screen cages, preferably of 1 ft³ (30 cm³) minimum size. Access is via a 24-inch (60 cm)-long cloth (surgical stockinette) sleeve that fits an opening on one side of the cage (see Figure 24.1). Adult mosquitoes may be transferred to or from cages by means of an aspirator, preferably a blower type, or a mechanically operated unit (Husbands, 1958). Adults may be fed on soaked raisins, honey, or table sugar. Three to 4 days after emergence, females may be fed on an immobilized chicken or laboratory animal, preferably by placing the animal on the mosquito cage instead of in the cage. Two to three days after the blood meal a shallow container of water should be placed in the adult mosquito cage for oviposition. Temperatures of 24–27°C and relative humidity of 70–80% are optimal for maintenance of *C. pipiens*.

Induced mating of *Aedes* species has been achieved with an anesthetization flask (Figure 24.2) developed by the Vector Biology Laboratory at the University of Notre Dame. Its use is described by T. N. Mather of the Entomology Department of the University of Wisconsin (personal communication).

Aspirate two to three 5-day-old female mosquitoes and anesthetize them with nitrogen. Place them over the hole in the vial cap forming the mating arena and introduce nitrogen slowly through the glass nipple of the apparatus. Several 4–6-day-old mosquitoes (genitalia must have rotated naturally) are aspirated into cotton. A male is impaled laterally through the thorax using a minute needle attached to a wooden applicator. The anesthetized female is tipped on her back over the air flow hole. Copulation is achieved by bringing the pinned male down at a 45–90° angle to the female so that the specimens are in the venter-to-venter position, and the tips of the abdomen are in proximity. Responsive males generally open their claspers, bending the tip of their abdomen toward the female genitalia when coupling is effected. Mating usually lasts 10–45 seconds. Males often mate more than once, and pinned males remain responsive for approximately 30–45 minutes. Best results are obtained at temperatures above 20°C and a relative humidity above 85%. Use of a stereoscope is advised in mosquito manipulation.

For details of mosquito culture beyond the introduction given here, see Gerberg (1970) and Belkin et al. (1965).

Flea rearing

Facilities and procedures for flea rearing vary greatly, depending upon the biological requirements of the species. One of us maintained thriving cultures of *Diamanus montanus* for months by merely placing the fleas in a 10-gallon ceramic crock furnished with a wire grid cover, a bottom layer of wood shavings, and a live ground squirrel, *Spermophilus beecheyi*. Food and water for the squirrel consisted of fresh carrots and laboratory rat chow. The squirrel was transferred to a fresh jar at 2- to 3-week intervals, and adult fleas were collected from the vacated crocks over a period of 1 to several weeks by means of an aspirator. Hudson (1958) describes a method for large-scale rearing of the cat flea that provides a 60-cm³ sheet metal box with a floor of hardware cloth and a subfloor consisting of a shallow metal funnel terminating in a glass jar (Figure 24.3). The cage is supplied with a shallow tray containing sawdust enriched with 2–5 parts of a mixture consisting of 100 parts by weight of ground dog food, 10 parts dried brewer's yeast, and 15 parts of dried beef blood. A flea-infested cat is maintained continuously in the cage, with food and water provided in trays attached to the inner walls of the cage. The larval medium is swept into the jar once a week, screened to remove coarse debris, enriched with more of the larval medium, and stored at 26.6°C and 80% relative humidity pending pupation of larvae.

The literature on rearing fleas of various species is extensive. A recent publication by F. G. A. M. Smit (1977) cites 45 of the more important references dealing with the subject over the period 1910–76.

Tick rearing

The literature on culture of soft and hard ticks is extensive. Gregson (1966) reviews general aspects of tick rearing and comments on rearing methods published for 28 species of ticks. He notes appropriately that "the life histories of different species of ticks contain so many peculiarities that, although the basic design is similar, rearing procedures must usually be fitted to the species under study."

Because all ticks are obligatory blood feeders at some times in their lives, availability of suitable vertebrate hosts is prerequisite to tick rearing. The suitability of a given avian, reptilian, amphibian, or mammalian species as a host varies, depending on the species and stage of tick in question. Most ticks require food after hatching and after each moult. The time required for feeding may be only a few minutes, as for nymphs and adults of many argasid ticks, or it may be 1 to several weeks, as for many ixodid ticks. Maintenance of free-living stages of ticks in general requires a relative humidity in the 70–90% range and temperatures above 22°C. Moulting of species such as *Ixodes scapularis* is best at 90–100% relative humidity at a temperature of about 26°C (32°C for larvae) (Harris, 1959).

Many types of infestation devices for feeding ticks have been described. For feeding nymphs or adults of fast-feeding argasids such as *Ornithodoros coriaceus* or *Ornithodoros hermsi,* the ticks may be applied to suckling laboratory mice in a plastic box closed with a screen cover; alternatively they may be applied to anesthetized laboratory animals held in suitable containers for 0.5–2 hours. For the slow feeders such as most ixodids, devices such as the feeding capsule (Figure 24.4) may be attached by tape to the clipped abdomen of the rabbit, guinea pig, or other suitable laboratory host and ticks introduced at will. Modifications of the device have been described (Gregson, 1966) suitable for use on cattle, dogs, and other large animals. Kohls (1937) used a cloth bag slipped over the ears of a rabbit and secured at the base to apply up to 30,000 larvae of *Dermacentor andersoni* or 80 adults of the same species. When small animals such as lizards and mice are used as hosts of slow-feeding ticks, the infested hosts may be held in small cages of hardware cloth supported over containers of water. Ticks are recovered from the water after they complete engorgement and detach. Larvae, nymphs, and adults of many ixodids survive at least 3 days' submersion in water, but larvae of some argasids will drown in a matter of hours.

Selected references on tick rearing in addition to those given by Gregson (1966) include Drummond et al. (1969), George (1971), Kaiser (1966).

Rearing other arthropods

Basic information on rearing other arthropods such as body lice, parasitic mites, *Culicoides* midges, black flies, stable flies, tsetse flies, screw-worm flies, bed bugs, and kissing bugs is reviewed in Smith (1966).

References

Belkin, J. N., C. L. Hogue, P. Galindo, T. H. G. Aitken, R. X. Schick, and W. A. Powder. 1965. Mosquito studies (Diptera, Culicidae). II. Methods for the collecting, rearing and preservation of mosquitoes. *Contrib. Am. Entomol. Inst. 1*(2):19–78.

Drummond, R. O., T. M. Whetstone, S. E. Ernst, and W. J. Gladney. 1969. Biology and colonization of the winter tick in the laboratory. *J. Ec. Entomol. 62*:235–8.

George, J. E. 1971. Drop off rhythms of engorged rabbit ticks

Haemaphysalis leporispalustris (Packard, 1896) (Acari: Ixodidae). *J. Med. Entomol.* 8:461–79.

Gerberg, E. J. 1970. *Manual for Mosquito Rearing and Experimental Techniques.* Bull. Am. Mosq. Control Assoc. No. 5. 109 pp.

Gregson, J. D. 1966. Ticks. In *Insect Colonization and Mass Production,* ed. C. N. Smith. Academic Press, New York, pp. 49–72.

Harris, R. L. 1959. The biology of the black-legged tick. *J. Kansas Entomol. Soc.* 32:61–8.

Hudson, B. W. 1958. A method for large-scale rearing of the cat flea *Ctenocephalides felis felis* (Bouche). *Bull. Wld. Hlth. Org.* 19:1126–9.

Husbands, R. C. 1958. An improved mechanical aspirator. *Calif. Vector Views* 5:72–3.

Kaiser, M. N. 1966. The subgenus *Persicargas* (Ixodoidea, Argasidae, *Argas*). The life cycle of *A.* (*P.*) *arboreus,* and a standardized rearing method for argasid ticks. *Ann. Entomol. Soc. Am.* 59:496–502.

Kohls, G. M. 1937. Tick rearing methods with special reference to the Rocky Mountain wood tick, *Dermacentor andersoni* Stiles. In *Culture Methods for Invertebrate Animals.* Cornell University Press (Comstock), Ithaca, N.Y., pp. 246–56.

Ryckman, R. E., and A. E. Ryckman. 1966. Reduviid bugs. In *Insect Colonization and Mass Production,* ed. C. N. Smith. Academic Press, New York, pp. 183–200.

Sawicki, R. M., and D. V. Holbrook. 1961. The rearing, handling and biology of house flies (*Musca domestica* L.) for assay of insecticides by the application of measured drops. *Pyrethrum Post* 6:3–18.

Schoof, H. F., 1964. Laboratory culture of *Musca, Fannia* and *Stomoxys. Bull. Wld. Hlth. Org.* 31:539–44.

Smit, F. G. A. M. 1977. *Flea News* No. 12. British Museum (Natural History) London.

Smith, C. N. (ed.). 1966. *Insect Colonization and Mass Production.* Academic Press, New York.

Smittle, B. J. 1966. Cockroaches. In *Insect Colonization and Mass Production,* ed. C. N. Smith. Academic Press, New York, pp. 227–40.

Spiller, D. 1966. House flies. In *Insect Colonization and Mass Production,* ed. C. N. Smith. Academic Press, New York, pp. 203–25.

Tarshis, J. B. 1961. Cockroaches for feeding purposes. *Zoonooz* 34:10–15.

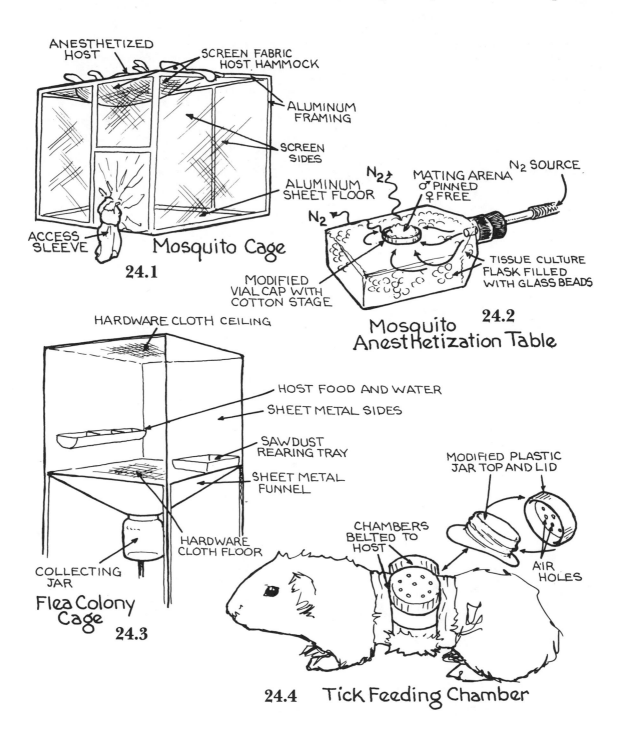

ANESTHETIZED HOST

SCREEN FABRIC HOST HAMMOCK

ALUMINUM FRAMING

SCREEN SIDES

ALUMINUM SHEET FLOOR

ACCESS SLEEVE

Mosquito Cage

24.1

N_2 SOURCE

MATING ARENA
♂ PINNED
♀ FREE

N_2

N_2

MODIFIED VIAL CAP WITH COTTON STAGE

TISSUE CULTURE FLASK FILLED WITH GLASS BEADS

Mosquito Anesthetization Table

24.2

HARDWARE CLOTH CEILING

HOST FOOD AND WATER

SHEET METAL SIDES

SAWDUST REARING TRAY

SHEET METAL FUNNEL

HARDWARE CLOTH FLOOR

COLLECTING JAR

Flea Colony Cage

24.3

MODIFIED PLASTIC JAR TOP AND LID

CHAMBERS BELTED TO HOST

AIR HOLES

24.4 Tick Feeding Chamber

CKKD

DATE DUE